Other Books by Gene Masters

Silent Warriors—The year is 1941. Shortly after the United States declares war on Japan in response to Pearl Harbor, Japan's Tripartite Treaty allies, Germany and Italy, declare war on America. The United States finds itself in a two-theater war. President Franklin Roosevelt sets as America's first priority the defeat of Nazi Germany, electing to wage a more-or-less holding war in the Pacific. In the beginning, the only force opposing the Japanese onslaught in the Pacific is the U.S. Submarine Service.

Jake Lawlor begins his war as executive officer aboard USS S-49, an aged S-class submarine, with orders to conduct unrestricted warfare against the enemy in the Pacific. When a freak mid-sea grounding causes the loss S-49, Jake assumes command of another boat, USS Orca, a new Gato-class submarine, under construction in Groton, CT. As Jake prepares a new boat and a freshly-assembled crew for war, the conflict in the Pacific is going badly for the Allies.

This is the story of Captain Lawlor's eleven war patrols, including an ongoing conflict with Imperial Japanese Navy Captain Hiriake Ito of the destroyer Atsukaze. The crew of the Orca is made up of grizzled veterans and wet-behind-the-ears youngsters, all working together for a single purpose: to bring an implacable enemy to its knees. Along the way, friendships are forged, and love affairs and marriages are created—and destroyed.

Operation Exodus—Six Christian missionaries are kidnapped by the Iranian government and become unwitting pawns in a high stakes gambit to embarrass a hated U.S. sitting president and ruin his chances of reelection. With the help of Israel's vaunted intelligence agency, Mossad, the prisoners are located in an ancient fortified prison just outside the southeastern Iranian port of Kanarak, a resort city off the Gulf of Oman. Rescue Operation Exodus is launched, headed by Navy SEAL Lieutenant Jake Lawlor, grandson of a legendary WWII submarine commander. Will the torture and brainwashing be successful, or will Jake and his fellow SEALs save the day? Find out in Operation Exodus.

The *Laconia* Incident

By

Gene Masters

The Laconia Incident
Copyright © Gene Masters, March , 2020

FIRST EDITION
10 9 8 7 6 5 4 3 2 1
ALL RIGHTS RESERVED
ISBN-13: 978-1-7346750-1-6

The names, places, and incidents are entirely historical — only a few of the characters are products of the author's imagination. Any resemblance to actual persons, living or dead is entirely coincidental.

In accordance with the U.S. Copyright Act of 1976, the scanning, uploading, and electronic sharing of any part of this book without the permission of the publisher constitute unlawful piracy and theft of the author's intellectual property.

This book may not be reproduced in print, electronically, or in any other format, without the express written permission of the author, except in the case of brief excerpts for publicity purposes.

Cover photo Cunard Line postcard
of the RMS Laconia circa 1921 - Public Domain
Periscope image © Shutterstock 105943382

Published in the United States of America by
Escarpment Press, Indian Land, SC

For Ruth
encore une fois

Introduction

This book is a dramatization of a true story. The incidents related actually happened, the ships described really existed, and the majority of the characters described in this book are real people.

The story of the *HMT Laconia,* and of her sinking by the German submarine *U-156,* has, of course, been told before, but, most hopefully, never better than in the pages that follow. There is, famously, a two-hour BBC series, also entitled *The* Laconia *Incident.* The series tells the same story, but it centers more around a fictional subplot than the incident itself.

In my telling of the *Laconia* story, the real events surrounding the sinking of the *Laconia* directly affect the lives of the people involved, and are, I believe, substantive enough to tell an amazing tale of both wartime savagery, political intrigue, and unexpected gallantry.

There have been other, nonfiction books written about the subject. One purely historical version, *The Sinking of the* Laconia *and the U-Boat War,* by James P. Duffy, proved an excellent resource. A second, *One Common Enemy,* by Jim McLoughlin, is a survivor's tale, and a great read, but it's more about McLoughlin's life experience, and his trials in an open lifeboat for twenty-eight days after the sinking. The *Laconia* sinking itself is a vital part of McLoughlin's story, but *only* a part of it.

My book relates the experiences of the fictional Robby Cotton, a Scots native who enlists in the British Royal Navy after his family is killed in a German air raid (the Clydebank Blitz of March 13-14, 1941). Seeking to avenge the death of his family, Robby enlists in the

Royal Navy, and is assigned to a gun crew aboard the battleship *HMS Victory*. He soon discovers that the ability to strike back at the enemy involves becoming exposed to risks that he hadn't planned upon: to be shot at, in turn, not only from the air, but also from the sea, and, most frighteningly, from under the sea. The danger from German U-boats is brought home when Robby witnesses the torpedoing and sinking of *Victory's* sister ship, *HMS Barham*. A series of subsequent events finds Robby serving as part of the gun crew aboard the *HMT Laconia*. And, only after having his ship literally sunk beneath his feet, does Robby discover that, after all, even his most dangerous enemy has a human face.

The Laconia *Incident* also tells the story of Marco Scarpetti, a sergeant in the Italian Army, who is captured by the British in North Africa, and that of one of his Polish guards, Stanislaw Kominsky. Both join Robby Cotton as passengers on the *Laconia's* final voyage, en route to Liverpool by way of Cape Town, South Africa, along with 1,800 Italian POWs, their Polish guards, and several hundred other British passengers and crew.

But *The* Laconia *Incident* revolves mostly about the true story of *Korvettenkapitan* Werner Hartenstein, captain of the German submarine *U-156*. It is Hartenstein's boat that attacks and sinks the *Laconia* on 12 September 1942. The subsequent, unprecedented actions taken by him, his crew, and the German U-boat command, after the actual torpedoing and sinking the *Laconia*, make a truly amazing tale. It's a story of how civility and mercy survive, even amidst the savagery and brutality of all-out war. It also shows how even the best-intentioned efforts can be foiled by stubborn adherence to well-established preconceptions, even in the face of overwhelming evidence to the contrary.

In this story, lives are both needlessly lost and gallantly saved. Mistakes, inscribed forever in the historical record, are made by the British Admiralty, the German High Command, and the United States Army Air Corps. And these tragic mistakes only unnecessarily added to the enormity of the loss of life originally engendered by this calamity. Yet, in contrast to the barbarity of modern warfare, the gallantry and humanity of one man, striving to do the right thing, stands out. This, then, is *The* Laconia *Incident.*

Gene Masters
Knoxville, Tennessee
2020

Prologue

12 September, 1942

Through his binoculars, *Korvettenkapitän* Werner Hartenstein studied the target from the bridge of *U-156*. The boat had just surfaced a half-hour earlier, and was now closing with the British transport that had been spotted the previous midday.

"What's her range now, Leo?" Hartenstein called down to his first watch officer through the open hatch to the boat's conning tower.

Inside the conning tower, first officer *Oberleutnant zur see* Leopold Schumacher, peering through the periscope, adjusted the range finder handle. He then read the numbers on the dial. "Target range is 5,100 meters, Captain," he sang out.

"Excellent," Hartenstein replied, his binoculars still fixed on the target. She was big, he noted, perhaps 20,000 tons, a gray ghost still inexplicably belching black smoke from her single funnel. And the weather was perfect for it: the sky cloudless, the air clear and still, the sea below black ink. There was practically no moon, he noted, just the sliver of a waxing crescent, yet the twilight sky was bright enough in these equatorial climes, and the visibility was really quite good. The outline of the ship stood out well enough against such a backdrop that Hartenstein could easily judge the angle her bow made to his line of sight.

"Right thirty degrees, I should think," Hartenstein mumbled aloud to himself. And the angle had drifted left just as one would expect from a ship on a steady course relative to his submarine.

"She's still not zigzagging, Captain," Schumacher called out. "Still steady on course three-zero-five."

"Very well, Leo. Let me know if she suddenly decides to change course."

"Aye, Captain," Schumacher replied, and then continued relaying periscope ranges and bearings to his fire control team in the conning tower and the control room below.

"I believe she's settled down for good, Captain, and for the night. She has not changed course nor speed since 1930 hours," Schumacher called up to the bridge a few minutes later. "Course is still three-zero-five. Speed is fourteen knots. If nothing else changes, we should be in firing range in just five minutes."

"Very good, Leo. Have the forward torpedo room prepare to fire three torpedoes—tubes one, two, and three."

"Aye, Captain," Schumacher replied, and relayed his commander's orders to the torpedomen forward. Hartenstein then ordered the U-boat to change course, keeping her bow pointed at the target, making her harder for her prey to spot, and a smaller target for the transport's gun should it come to that. A single-barrel, large-caliber, naval gun could just barely be made out, mounted on the ship's stern, and Hartenstein was quite reasonably wary about coming under that weapon. As an added bonus, keeping the bow pointed at the target also minimized the torpedo deflection angle required for a hit.

Five minutes passed. The target's bow angle had settled out at right ninety degrees.

Perfect, Hartenstein thought.

"Range now, Leo?" he asked.

"Target range eight hundred meters, Captain."

"Good. Do you have a firing solution?"

"Yes, Captain. Recommend deflection angle eight degrees right."

"Very good, Leo. Send the order forward, set deflection angle eight degrees right, set torpedo depth four meters."

"Aye, Captain." Then, sending to the torpedomen forward, "Tubes one, two, and three, set deflection angle eight degrees right, depth four meters."

"Fire one, Leo." Schumacher relayed the order.

Even though the boat was making way forward at seven knots, the pulse of the fired torpedo could be felt through the pressure hull. Thirty seconds later, the procedure was repeated, and a second deflection shot was fired at the target. Hartenstein decided to hold off on the order to fire the third torpedo.

The first torpedo struck His Majesty's Transport *Laconia* at 8:07 PM local time. The torpedo struck forward of amidships, and the ship first listed sharply to port from the force of the explosion, and then rolled back to starboard.

The second torpedo struck thirty seconds later, just outside number two hold, pitching the ship violently back again to port.

Only minutes later, *HMT Laconia's* Captain, Rudolph Sharp, foreseeing the inevitable, ordered his third officer, Thomas Buckingham, to pass the order to abandon ship. Buckingham broadcasted the message throughout the ship: "Abandon Ship! Abandon Ship! All passengers will proceed in good order to their assigned lifeboat stations. Abandon Ship!"

Hartenstein ordered *U-156* submerged, wary lest the target's gun be manned and start firing on her attacker. But the *Laconia's* severe list to port precluded any possibility of her crew training—never mind firing—the gun.

There was sufficient time—just over an hour—from the first torpedo strike and her ultimate demise, for an indeterminate number of the ship's lifeboats to be launched, each crammed full of survivors. Many of the boats couldn't be launched at all, because of the ship's severe list. So it was that many more souls simply jumped over the side, braving the open water and the inevitable presence of sharks and barracuda.

At 9:11 PM, 12 September 1942, her boilers exploding, the *Laconia* sank stern first.

Of the 2,732 souls aboard the *Laconia* that night, only 1,113 ultimately survived. And, had it not been for the heroic efforts of the captain and crew of the U-boat that had torpedoed her in the first place, the death toll—1,619 souls—would have been far, far, worse.

Part I

Prelude to Calamity

1

The Clydebank Blitz, March, 1941

"Slow down, Robby," his mate, Harlan, told Robby Cotton, in a thick Scots accent. "Glasgow ain't about to run out of brew anytime soon! No need to try and drink it all tonight!" Harlan's blue eyes twinkled merrily, as he gleefully chided his friend. Truth be told, Harlan White himself had consumed quite as much as Robby had. But then, he was half-again Robby's size, tall and muscular, and could hold his liquor better.

"Sod off, Harlan, you're just crying 'cause you can't keep up, is all," Robby replied untruthfully through a beery haze. It was not yet ten o'clock that Thursday in March, and Robby and his mates were already three sheets to the wind. "'Sides, it ain't every night your old schoolmates go off and join the Navy! No tellin' when I'll ever see your sorry arses again," Robby added.

"You could've joined up as well, you know," his other mate, Donald Conklin, joined in, "But we'll probably blast them Jerries off the sea before you get a chance to come aboard and fire a shot!" For whatever reason, Donald attempted to stand up, but, unlike Harlan, the weight of the drink was too much for his slight frame, and he slunk back down into his seat again.

"No worries, there, Donald, no worries there." Robby replied. The beer had slowed his thinking, and it took a few seconds before he added, "Plenty of Jerries left before I get drafted. I'll be eighteen soon enough, lots of time to get shot at then."

"You mean your ma wouldn't sign for ya, don't you Robby?" Harlan taunted. Like Robby, both Harlan and Donald were just seventeen, and their parents had signed the papers for them to join the Navy as minors.

"You sod off, too," Robby replied, paused, and then added, "that may be true, all right, but 'tis also true I ain't at all in a rush to get shot at, mate." He looked down and realized he had drained the mug in his hand. "But enough of this. You mates are boardin' a train, and will be bound for Plymouth tomorrow at noon, and there's plenty of drinking and carrying on to do afore then!" He called out to the oversized waitress a few tables away, "More beer here, my darlin'. Me mug is empty. More beer here!"

* * * * *

The newspapers had called it the "Clydebank Blitz." It was a series of two Luftwaffe raids on the night of 13 March, and the early morning of 14 March, 1941 over the towns of Clydebank, and nearby Dalmuir, some eight miles downriver from Glasgow. The targets had been the shipbuilding works on the River Clyde and the nearby munitions plant.

Less than ten percent of the bombs dropped had hit their targets. The munitions plant was completely destroyed. The shipyard where the keel of the *Tyrrhenia* had been laid down twenty-two years earlier, however, escaped with only minor damage.

But the errant ninety percent of the bombs were not without their deadly effect. The surrounding townspeople suffered terribly. Some locals, deep asleep in their beds, never woke, this despite the shrill cry of the air raid sirens and the drone of the bombers overhead. It was probably just as well. At least, when the bombs stuck their homes, most never felt a thing.

Just as they had been trained, the home guard manned the gun emplacements when the sirens sounded. But, while the civilian gunners had practiced firing live ammunition, none had ever fired at an actual target. Despite the determined enthusiasm of the home guard, and a full moon and clear skies, none of the anti-aircraft guns scored a single kill.

The butcher's bill included some 1,200 civilian dead, another 1,000 seriously injured, and some 12,000 homes destroyed. The Germans, in turn, had lost just two aircraft in the raids, and those to RAF Hurricane fighters.

The United Kingdom had already been at war with the Axis Powers for just over eighteen months. But, until that March night, for the people in southern Scotland, the fighting war was the home guard playing soldier, and something you just read about in the newspapers.

Among the homes destroyed in the raid was the Dalmuir home of Howard and Mary Cotton. Both Howard and Mary and their three daughters were asleep inside the house when the bomb struck, and all five were killed. Their eldest son, Robby, was not at home at the time. Robby was in Glasgow on the night of the 13th and 14th, just eight miles away. He would have seen the flashes and heard the bombs going off in the distance, had he only been watching and listening.

Gene Masters

* * * * *

A sweaty Robby Cotton was lying naked atop the covers when he awoke that Friday morning, 14 March. He had the urgent need to piss. Sunlight filtered through the dusty window at the foot of the bed, and the room was stuffy and hot. Sometime during the night, the electric fire must have been turned way up to counter the chill of the Glasgow night. As it was, and in contrast to the inside temperature, the hazy morning, outside air was just above freezing. Robby's head was pounding, the hangover a result of all the alcohol he had consumed the evening and night before. His mouth was rank and dry, his every tooth feeling like it had been swathed in cotton nappies—and his aching bladder was full.

Only when he finally opened his eyes wide enough did Robby become aware of the woman sleeping next to him. In contrast to his own spare, lean body, this woman's naked body was large, with wide hips and thick legs. She was lying on her side, facing him, and like him, had slept atop the covers. He looked at her hard, but, try as he might, he could not remember her from the night before. All he could recall was that Harlan, Donald, and he were drinking and having a grand old time. Yet, there she was. *She's got to be old*, he thought, *she looks like she's really old. At least thirty.*

Her tousled hair was black and dull, her face in repose peaceful, but unremarkable. Her breasts were large in keeping with the rest of her, nestled one atop the other, right upon left, two brown nipples set in tan areolas—two owl's eyes staring at him. What struck Robby as truly remarkable was her skin. It was flawless, the color of clotted cream, and totally unblemished. Without thinking, Robby reached out, and, with the back of his hand, stroked the flawless skin just above the woman's hip. It felt every bit as good as it looked: soft and warm—and yielding.

Upon feeling his touch, the woman sighed, and turned over on her back, the owl flown.

Mustn't wake her, Robby thought, and, remembering again his urgent need to piss, eased himself off the bed. He searched the room for the chamber pot, and finally found it in a cupboard in the near corner of the room. The cupboard open, he relieved himself into the chamber pot. His flow made some noise, but the woman slept on.

He found his clothes among hers, scattered on the floor of what must have been the woman's flat. Once dressed, he tiptoed out of the flat, closed the door quietly behind him, and made his way down four flights of stairs and into the street below. The icy air greeting him sucked the breath from his lungs, and make his head pound ever harder.

Robby had no idea where he was, but thought he'd best find someplace to get some aspirin, and perhaps, if he could hold it down, some breakfast. He thought then about Harlan and Donald, wondering where they were. *But they don't have to catch their train 'til noon*, he mused *So they'll probably be just fine.*

He started down the street, his back to the not yet risen sun, the faint light fighting hard to bite through the city haze. It was at the street corner where Robby saw the headlines at the newsstand there. In utter disbelief, he bought the paper, the *Daily Record and Mail*, and read about the nighttime bombings. His home had to be in the thick of it.

Now all he could think about was getting home.

*****

Hard times had fallen on all of Scotland with the Great Depression, but had passed over the Cottons. Howard had steady work throughout those lean years as a steelworker at the William Beardmore and Company Shipbuilding Works in Dalmuir, on the River Clyde. The Cotton family, while hardly living high on the hog, had never suffered economic hardship during those prewar years. The Cottons always had food on the table and a roof over their heads—and there was even just enough money to send their children to the Catholic school. Now, with the war, and with Howard taking a supervisory job at the Beardmore works, things were actually looking up.

The only sticking point in Mary and Harry Cotton's marriage was their eldest child and only son, Robby. Robert Miles Cotton was born five years after his father began work at the shipyard. That was on 17 January, 1924.

Robby had been a difficult child, strong-willed and rebellious. He preferred roaming the Clyde River estuaries downstream of Dalmuir with his mates Harlan and Donald, all three playing hooky from St. Stephens School *and* the demanding nuns who taught there. And when he did go to school, it seemed that Robby was always getting into scraps with his schoolmates. He got along well only with the likes of Harlan White and Donald Conklin, ne'er-do-wells just like himself.

"There's a job for you at the Works, Robby," Howard had told his now 17-year-old son, "a good one—apprentice ship fitter. Pay's good, even to start. You'll learn a trade, and with a job like that—working for the national defense—you'll be kept out of the shootin.'"

But Robby Cotton wasn't interested in working, and though his mates were set on joining up, Robby was not. And an indulgent Mary Cotton slipped enough spending money to her beloved only son, that he could afford to defy his father, at least for the time being. When he turned eighteen, he figured, then they'd come after him for the Army or whatever, and then he'd worry about the national service. Maybe then he'd sign up for the Navy, like Harlan and Donald. Besides, that wasn't for months yet, and, in the meantime, Robby intended to enjoy life. However, the events of 13 and 14 March did for Robby what his father's efforts could not. Robby Cotton suddenly grew up.

* * * * *

The recruiter, a ruddy-faced, portly chief petty officer, addressed the lean, blond, likely-looking young man that had shown an interest in joining the Royal Navy. "Well, if it's shootin' at Jerry airplanes you're wantin', then the Navy's the only place for you, Lad," the Chief assured Robby. "You'll be trained to shoot down Jerry planes just as soon as you're sent down to a ship. That's what the Navy does with likely lads like yourself!"

"Yes, Sir," Robby explained, "it was them Jerry airplanes that murdered my family asleep in their beds, and all I want's a chance to shoot back at the bastards."

"Don't 'Sir' me, boy, I'm just a petty officer, but you can call me 'Chief,' as is proper. Now when did you say you was born?"

"It was 17 January, 1924, Chief."

"You say you was born on 17 January . . . *1923*?" the chief corrected.

"No, Chief, *1924*," Robby said to the recruiter, who then scowled back at him. "17 January . . . *1924*," Robby repeated for emphasis.

"Now you see, m'boy, there's this," the chief explained patiently, just as Robby had seen the nuns do at times, "If you was born in 1924, you'd be just seventeen, and you'd be too young to enlist without your parents' permission—and which you can't really get now, after all, with them being killed by the bloody Jerries and all . . ."

"Ah, yeah," Robby replied, as the light dawned, "you heard right, Chief—1923 it is!"

"That's what I thought you said," the chief replied.

With that, Robby Cotton signed up in the Royal Navy for "the duration of hostilities."

Within the week, Ordinary Seaman Robert Miles Cotton reported aboard the Royal Navy Barracks in Plymouth, *HMS Drake*, for training. For whatever reason, Robby discovered, the Royal Navy considered shore installations "stone frigates."

2

Nine Months Earlier, HMT Lancastria, *June, 1940*

A full two weeks after the British had extracted their forces from the beaches of Dunkirk, on 16 June, 1940, Royal Merchant Navy Captain Rudolph Sharp, master of the troopship *HMT Lancastria,* anchored his ship some eleven miles southwest of St. Nazaire, in the Loire River Estuary, in what was, technically, German-occupied France. Sharp and *Lancastria* had been dispatched to St. Nazaire to take part in Operation Ariel, the evacuation of British and Allied forces and nationals from western France.

The merchant ship, *RMS Lancastria,* a Cunard-White Star Line ocean liner, had been requisitioned by the British government in April, 1940, and was hastily converted for use as a troopship. The 578-foot-long ship, displacing 16,200 gross tons, was originally commissioned *RMS Tyrrhenia,* and had been built in Dalmuir, Scotland, by William Beardmore and Company in 1920.

The *Tyrrhenia* was the first vessel on which Howard Cotton, a newly-hired apprentice steelworker at the Dalmuir Shipyard, had ever worked. Howard Cotton's only son, Robby Cotton, and Rudolph Sharp were destined to have their fates intertwined, even though they would never actually engage in face-to-face conversation.

In 1924, *Tyrrhenia* was refitted and rechristened *Lancastria*, and placed in service on the regular Cunard-White Star route between Liverpool and New York. Now, in St. Nazaire, instead of well-heeled transatlantic passengers, *Lancastria* was tasked with transporting refugees from the Nazi juggernaut that had swept across France in just forty-six days.

Once at anchor in the estuary, *Lancastria* began to take on refugees as soon as they could be ferried out to the anchorage. The ship's official capacity was 2,200 passengers and crew, but Sharp had been ordered to disregard international law and load aboard "as many men as possible."

It was a clear night with a waxing full moon, and the visibility was excellent. The inky harbor waters were quiet, with only gentle swells. Boats small and large moved back and forth, carrying passengers from the docks throughout the night. Before morning on the 17th, the ship was already loaded with human cargo well beyond her rated capacity. And still the boats shuffled back and forth from the port.

The Luftwaffe began attacking the shipping assembled for Operation Ariel on the early afternoon of 17 June. It was, unfortunately, perfect flying weather: clusters of billowy white clouds high against a clear, brilliant blue sky.

The Germans had quickly set a nearby vessel, *HMT Oronsay*, aflame, when Captain Sharp was given leave to get *Lancastria* underway for England. *Lancastria* was now loaded with an uncertain number of passengers, later estimated at somewhere well above 4,000 souls, possibly even as many as twice that number. But after the enemy planes had departed, Sharp, wary of enemy submarines, elected to stay put and wait for a promised destroyer escort rather than getting underway immediately. That soon proved to be an unfortunate decision.

Later that same afternoon, a second wave of Luftwaffe Ju 88 fighter bombers hit the port.

The first bomb to hit *Lancastria* struck at 3:48 PM. The ship immediately began listing to port. The first hit was followed closely by two more. Captain Sharp ordered the ship abandoned—the order superfluous, since a panicky, undisciplined, crew had already begun lowering lifeboats. Not long afterward, a fourth bomb went down the ship's funnel, exploded in her engine room, and released about 1,200 tons of fuel oil into the estuary.

His ship broken apart and aflame, Sharp did his best to continue supervising an orderly abandoning of the ship, while strafing attacks by German aircraft, directed at the swimming survivors, set the floating fuel alight. Survivors not killed by the strafing were either burned alive or choked to death by the fumes. In the end, of the passengers who were aboard *Lancastria,* only 2,477 survived. Among them, and the last living being to leave his ship, was Captain Rudolph Sharp. In the fog of war, *Lancastria's* exact death toll was never determined.

Gene Masters

3

HMS Valiant, *October, 1941*

Seven months after the Clydebank Blitz, on 23 October, 1941, ordinary seaman Robby Cotton reported for duty aboard the battleship *HMS Valiant*, in Malta.

"Ordinary seaman's mess is three decks down, port side, aft. Find yourself an open spot and hang your hammock there, for now," the petty officer on the quarterdeck told Robby, after he had presented his draft chit.

Lugging his hammock and duffel bag, Robby strode across the main deck to the port side, and headed aft, looking for a passageway that headed belowdecks.

The ship was huge, 33,000 tons displacement, and when he had learned she was to be his new home, Robby had gone to *HMS Drake's* library, looked her up, and memorized her particulars. *Valiant* was 644 feet long, and 91 feet across. And there were sailors everywhere—everyone headed somewhere on some very obviously important mission. Robby gazed up and marveled at the size and number of *Valiant's* guns. Again, the memorized statistics were recalled: eight 15-inch guns, fourteen 6-inch, and two 3-inch anti-aircraft mounts. *It's them AA guns I'll be wantin' to shoot,* Robby mused, *if only they'll let me!*

Making his way aft, Robby had to work his way around two torpedo tubes (he knew there were two more just like them on the starboard side). Gazing up and about, he marveled at the height and complexity of her superstructure, the maze of doors and passageways aboard *Valiant*.

Robby located a ladder (actually a staircase, but aboard His Majesty's ships stairs are called "ladders," just as walls are called "bulkheads") and made his way down into the bowels of the great warship. Counting the main deck as the first deck, he went down two more decks and found a compartment with hammocks strung throughout. There, as the petty officer on the quarterdeck had instructed, he found an open spot and strung up his hammock.

"Hello, Mate, welcome aboard."

Robby looked for the source of the greeting.

"Here, man, over here."

Robby spied a prone form, lying partially hidden, two hammocks inboard. "Hello! I'm Robby Cotton, and you?"

In a practiced maneuver, a smiling figure rolled out of the hammock. The man's smile lit the place up. He had a shock of wavy brown hair, bright brown eyes, straight nose, and a lantern jaw—overall, Robby thought, quite a good-looking fellow.

"Able Seaman Jim McLoughlin. I hear a Scots brogue. You wouldn't be a Scotsman, now, would you, lad?"

Robby laughed. He'd been regularly teased about his accent during training at *HMS Drake*. "Born and bred I am," he replied, "from just outside Glasgow."

"Right, Minnow," McLoughlin said, spying the rating patch on Robby's arm. "I'm from Liverpool, myself. But here, lad, you're in the wrong place. This here's the Able Seamen's quarters. Ordinary Seamen are on the deck below."

"The Petty Officer on the quarterdeck said 'port side, aft, three decks down,'" Robby pleaded in his defense.

"Right, that," McLoughlin allowed, "but you're only *two* decks down. This here's the third deck, but it's only two decks down from the main deck. You'll be wanting the fourth deck."

"I think the Navy arranges things just so's they can drive a man potty!" Robby exclaimed.

McLoughlin laughed. "C'mon, Minnow, I'll get you settled," he offered, still smiling. He then led Robby down one level to the ordinary seaman's mess, and found him an open spot to string up his hammock.

* * * * *

Robby's first duty station outside the UK was not at all what he had expected. When he reported aboard, nobody had bothered to tell him that the Italians bombed Malta practically every afternoon. He discovered that shooting at enemy aircraft involved being their target in return, and while his battle station was to his liking—the port-side mount 31, one of the ship's two three-inch AA guns—he was just a loader. *If I'm gonna get shot at*, he thought, *I at least want to be aiming the bloody gun!*

He was happy to discover it was also Tom McLoughlin's battle station, and it was McLoughlin, also a loader, who showed him how to hustle a shell from the ready locker and into the gun breech, just in time for it to be rammed home and fired—without losing a finger or two in the process.

"You're every bit as important to the running of the gun, and shooting down of the enemy," McLoughlin had assured him, "as the bloke what aims the gun, Robby. It's the teamwork that shoots down them Dago bastards." (Not that Robby particularly wanted to shoot at Italians—it was Germans, after all, who had killed his family—but until given the opportunity to shoot at German planes, their Italian allies would have to do.)

During their stay at Malta, Robby's gun crew managed to shoot down two Italian Fiat BR.20 bombers, and take partial credit for a third, since one of the six-incher crews also claimed that kill. Robby felt then that he had somewhat evened the score for what the Germans had done to his family, even if just a little.

With such regular afternoon practice, Robby soon became a proficient loader, almost the equal of his mentor, Able Seaman McLoughlin. Despite the disparity in age (McLoughlin was almost twenty) and in their ratings, the two men quickly became friends.

Robby made other fiends as well. One of them was Ralph Tinsdale, a mate from *HMS Drake*, who had reported aboard *Valiant* a day after Robby. He hardly knew Tinsdale while at *Drake*, but Ralph's freckled face was at least a familiar face, and the two had sought each other out. Ralph was from Bath, had a slight build much like Robby's, and was the first in his family to serve in the Royal Navy. Besides McLoughlin and Tinsdale, Robby's immediate circle of friends eventually included three other ordinary seamen: Charles Martin, James Fellows, and Arthur Kinkaid.

The Laconia *Incident*

Charlie Martin was from Harrow, heavy set, and like Ralph Tinsdale, a redhead. James Fellows was sallow-skinned, tall, and gaunt. He was from Dover. Arthur Kinkaid was a lean Londoner, with an athletic build, fair skin, and jet-black hair. Counting Robby, all five ordinary seamen looked up to Jim McLoughlin, somewhat in awe of his advanced skills (for them) and his abilities.

Also helping to bond the five of them was that they were on the same watch-stander list, and were thus able to go ashore for liberty together—not that Malta was all that great a port for liberty. Aside from those residing on the four streets immediately adjoining the base, the native Maltese were not all that friendly toward their British defenders. But those four streets had all a young sailor could ask for: food, booze, female company, and rooms that conveniently rented out by the hour. For those unwilling to waste time wining and dining one of the local Maltese lovelies, there was also the choice of either of two brothels.

First, aboard *Drake,* and later reinforced at quarters aboard *Valiant,* Robby and his mates had been thoroughly schooled in the probable dangers to their health and well-being, literally embodied by the available local females. That message had not been lost on Robby and Ralph, and their mates, Charlie and James. But Arthur considered himself a ladies' man, and cheerfully gave in to whatever temptations presented themselves. And those always took the form of both booze and women, and always in that order. They had also been schooled in the necessary precautions to be taken should they fall prey to temptation, but Arthur had apparently missed, or had slept through, that portion of the lecture.

At home, and especially in Glasgow, Robby (despite having been born a Catholic, and knowing it was a mortal sin) was only too eager to bed any girl willing to have him. But the vivid posters, and the grainy black-and-white films that the Navy showed all their young recruits, had put the fear of God into him. So Robby decided it was perhaps time to take his religion seriously, avoid sin, and confine himself to just the beer. And so, also, and for whatever other reasons beyond fear of venereal disease, did all his other mates. Or they did, at least, for the brief time they were in Malta. All of them, that is, excepting Arthur.

Between the afternoon air raids and the quality of Maltese liberty, Robby was understandably happy to hear the news that *Valiant* was leaving Malta for Crete. The Germans had invaded Greece and were threatening Crete, and *Valiant* was getting underway en route to Suda Bay, Crete. Now, he reckoned, perhaps air raids would not be a daily occurrence.

"Well," Robby said to McLoughlin on hearing the news, just after they had stood down from yet another Italian air raid, "at least now maybe we won't be getting shot at every day."

McLoughlin laughed. "Hold that thought, Minnow. I think you may be in for a bit of disappointment."

"What do you mean?"

"You'll see soon enough."

The Laconia *Incident*

* * * * *

In Suda Bay, true to McLoughlin's prediction, Robby was indeed disappointed. He again saw air action, and this far more intense than at Malta. But this time, at least, they were firing at the invading Luftwaffe. *At last,* Robby thought, *I get to shoot at Germans!*

Robby got his chance to shoot at Germans, all right. In Malta, one could almost set his watch by the Italian air raids. Weather permitting, the Royal Italian Air Force could be depended upon to show up between one o'clock and one-thirty in the afternoon, and to depart the area forty-five minutes after arrival. And the Italians never attacked on a Sunday.

The Luftwaffe in Crete held to no such schedule. Air raids could be expected at any time of the day or night, and Robby spent many a night dozing off while at his station on mount 31, only to be nudged awake by Jim McLoughlin, or another of his mates.

During two of those nights, at least, Robby got a smell of victory. In the wee hours of Thursday, 30 October, and again, on Wednesday, 5 November, mount 31 was given credit for splashing two Focke-Wulf 200 "Condor" bombers.

"At last," Robby crowed to McLoughlin and the others, "we killed us some Germans, we did! We showed them Nazi bastards!"

"No need to go all cock-a-hoop, Robby," McLoughlin chided. "Plenty more of those buggers got through, dropped their bomb loads, and returned to base without a scratch. And they'll be comin' back tomorrow, and again the next day, and the next. They do their bit and we do ours. Just goes on and on."

But Robby refused to have his enthusiasm dampened.

Because the air raids could be expected at any time of the day or night, there was no liberty in Suda Bay whatsoever.

Valiant stayed moored in that port for what was, for Robby and his mates, a busy and sleepless seventeen days before retiring to safe haven: Alexandria in Egypt. Robby's reaction to the move out of Suda Bay was mixed. He was happy to be standing down for a while, and not getting shot at, but he had in no way sated his need for revenge for what the Luftwaffe had done to his family at Dalmuir. He had, after all, joined the Navy to kill, and get even with, Germans.

4

HMS Barham, November, 1941

Valiant **got underway from Alexandria in company with the battleships *HMS Queen Elizabeth* and *HMS Barham,* and a screen of cruisers and destroyers, to join up with a larger fleet.** It was the last week of November, 1941. Their mission: to intercept two Italian convoys en route from Messina, Italy, to Benghazi, Libya, in Italian East Africa.

The Italians had invaded Ethiopia in 1935, and had defeated and annexed the country into Italian East Africa in 1937. Now, the British ground forces, sweeping west out of Egypt, had driven the Italians back to Benghazi. The beleaguered Italian army was desperately short of supplies, hence, the convoys out of Messina en route to provide succor to their hemmed-in forces.

Now, a British battle fleet combed the Mediterranean in an effort to locate, intercept, and destroy the Italian supply ships.

* * * * *

Oberleutnant zur see Hans-Diedrich von Tiesenhausen, in command of the type VIIC German submarine *U-331,* couldn't believe his luck. Just over an hour earlier, while cruising on the surface, his lookouts had spotted a number of masts, hull down, on the horizon. But as he drove his boat forward, and the masts drew closer, he thought his luck had left him. The way the masts were moving, their bearing shifting sharply left, there was no way he could get his boat in a position to attack.

"They're moving away from us," von Tiesenhausen observed to his first officer. "Damn!"

"Perhaps not, *Kapitän*," his first officer observed. "The bearing drift appears to have slowed. They are zigging, either away, or toward. Time will tell. If the masts draw closer, and they have zigged toward us, we may still be in luck."

And draw closer, they did.

By the time the ships heading toward *U-331* were identifiable, it was already dusk. But, despite the failing light, it was clear to von Tiesenhausen, that a British battle fleet was bearing down upon him. Even after nightfall, the oncoming destroyers were clearly illuminated by the first-quarter moon behind them, bright in the clear, virtually cloudless, night sky. And the British had not yet spotted the low silhouette of *U-331* in the slate-colored, choppy waters before them.

Von Tiesenhausen then quickly submerged his boat, and, showing his periscope for only seconds at a time for quick observations, deftly maneuvered *U-331*, penetrating the screen of destroyers and cruisers undetected. It was a battleship he wanted, and it was a battleship he would sink! "Prepare all tubes forward," he ordered.

And there was a beauty in his crosshairs. The battlewagon was moving in front of him now, its bow about sixty degrees to the left of his line of sight, moving from right to left. "We will fire all tubes forward, make

The Laconia *Incident*

ready tubes one through four," Von Tiesenhausen ordered. He then gave direction to his first officer. "The first shot will be a deflection shot from tube one, angle seven degrees left, set depth four meters."

U-331 carried a full complement of fourteen G7e electric torpedoes, each tipped with 280 kilograms of high explosive. The G7e was standard throughout the U-boat fleet. It was silent, and, unlike the American steam-powered torpedoes, left no wake.

Seconds later, the forward torpedo room reported back to the conning tower. "Tube one ready, deflection angle set seven left, depth set four meters."

"Very well. Standby—fire one!"

In the forward torpedo room, the torpedoman pressed the firing button, and an electric torpedo was quickly on its way to the target. It was followed in quick succession by three other torpedoes, three more deflection shots.

Soon a flash was seen through the periscope, and, a split second later, the explosion heard. But there was little time to gloat, or to savor the moment. *U-331*, now very light forward with all four forward torpedo tubes emptied, broached the surface, the boat's bow, periscope, and bridge superstructure suddenly exposed to the enemy. Before the British could react, however, Von Tiesenhausen recovered control of the boat, ordering "all hands forward," flooding trim tanks, and driving the boat deep, with the control planes on full dive, and at full battery power. As these maneuvers were still in progress, two more explosions were heard—then, nothing; the fourth shot was apparently a miss.

Once the ship was deep, Von Tiesenhausen ordered *U-331* into a tight right turn. Then *U-331* was slowed, and crept silently away to the east. Except for the unplanned broach, Von Tiesenhausen's boat had been unseen and undetected by the British throughout the entire action. Now he had only to once again evade the screen and make for open water.

* * * * *

It was 26 November, and the formation was still searching for the Italian supply convoys. Robby was on his watch station: forward port lookout, main deck. Suddenly, he was startled by a blinding flash to port, and, simultaneously heard the first explosion. He was sure that the *Valiant* had been hit—and certain he would soon be swimming in the Mediterranean.

But it was not *Valiant* that was hit. Two more explosions followed, and Robby saw that the second and third flashes to port were across the water, and the *Barham* was the source of all three explosions. He was horrified to see the havoc wrought so quickly as the three torpedoes ripped apart the dreadnaught. The mighty *Barham* suddenly rolled over, and, just as quickly, her boilers and magazines exploded. In under ten minutes, she disappeared completely. *HMS Barham* took 860 men down with her to the bottom of the sea, in far too little time for Robby to process the tragic events that had unfolded before him.

Later, two destroyers managed to fish another 489 men from the water—those who were able to clear the ship before it sank. Two of them later died from their wounds.

U-331, the U-boat that had penetrated the screen and sunk the *Barham*, got away clean.

The incident so rattled Robby that he didn't sleep well for the rest of the deployment. There had been little enough opportunity to sleep on any deployment to begin with, and by the time Robby heard that the Italians had evaded them after all, and that *Valiant* was to return to Alexandria, he was worn to a frazzle.

Time and time again he had pictured *Valiant* falling to the same fate as *Barham*. German torpedoes were electric, and left no wake. His ship could be struck at any moment, and no lookout, however astute, would ever see a torpedo coming. On watch, as a lookout, he was supposed to scan both the sky for aircraft, and the sea for periscopes. Robby, without even realizing that he was doing so, now gave short shrift to the former, and concentrated heavily on the latter, searching the sea for periscopes.

You can see the bloody planes coming at you, after all, he thought, *and shoot back at them. But there's no tellin' when a bastard Nazi sub's got you in its sights — there's no tellin', and there's no defense!"*

On 17 December, *Valiant,* in company with *Queen Elizabeth*, and the destroyer *HMS Jurvis,* returned a now battle-scarred and battle-weary Robby Cotton to safe harbor in Alexandria.

Gene Masters

5

Alexandria, December, 1941

Luigi de la Penne's fellow frogmen in the *Decima Flottiglia* MAS teased him, saying that the tall and lean de la Penne looked just like the American singer, Bing Crosby. He had heard it all before. Crosby's movies and recordings had been popular in Italy for at least a decade, and the teasing had begun during de la Penne's tenure at the *Accademia Navale*, the Italian Naval Academy in Livorno, when his classmates had first noticed the resemblance. He pretended to shake it all off, but was secretly pleased — Crosby, after all, was at least as well known around the world as *Il Duce* himself.

There was a new moon on the night of 18 December, 1941, but the air was clear, and a southern breeze, coming up from Africa, pushed whatever wispy cloud cover there was out and over the Mediterranean. From the deck of the surfaced Italian submarine *Scrire, Tenente di Vascello* (Lieutenant) Luigi de la Penne, and his team of six frogmen, observed three important units of the British Mediterranean fleet as they entered Alexandria harbor.

Luigi watched intently as the two battleships and a destroyer entered the anchorage. These, he knew, were *HMS Valiant* and *HMS Queen Elizabeth*, along with the destroyer *HMS Jurvis*. The three vessels anchored close aboard an oiler, *HMS Sagona*, and two smaller auxiliaries. The British Admiralty deemed Alexandria a safe harbor, well out of reach of Italian Navy Units and German submarines. Now, Luigi hoped to shake that confidence.

Launching from *Scrire*, de la Penne, his second, the buff and fit *Sottotenente di Vascello* (Lieutenant Junior Grade) Emilio Bianchi, and four others, drove three, two-man underwater assault vehicles into the harbor. The vehicles were disdainfully referred to by their operators as *"maiali,"* or "pigs," because they were notoriously difficult to maneuver.

The six divers each used an oxygen rebreather apparatus. These devices, originally developed by the British, used carbon dioxide absorbent canisters and oxygen bottles to supply breathable air underwater. This they did without expelling any gasses; thus there were no telltale air bubbles to rise to the surface.

Struggling with their pigs, the frogmen managed to penetrate the harbor defenses and attach limpet mines to the battleships, the destroyer, and the oiler. It was de la Penne himself who attached the mine to the keel of the *Valiant*.

But then things went very wrong for de la Penne. His mask began to leak badly, causing his rebreather to malfunction. There was nothing he could do—he could not breathe—and de la Penne was forced to the surface. Bianchi refused to leave his leader, and surfaced alongside him.

* * * * *

It was just toward the end of the evening watch. Robby Cotton was standing the watch as roving guard on the main deck, when he heard what sounded like the bursting of air bubbles, followed by the splashing of water, just off the port side of the prow and in the water below.

"Who goes there?" Robby shouted down from the deck of the *Valiant*, shining his flashlight onto the two men bobbing on the surface not ten feet off the battleship's port bow. Without hesitation, Robby sounded the alarm: "Divers in the water! Divers in the water!"

As their four companions made their way back safely to the *Scrire*, and as their own discarded *maiale* settled into the harbor mud below them, de la Penne and Bianchi were hauled aboard one of *Valiant's* motor launches. They were soon aboard the battleship.

The two men were brought to the ship's captain, Royal Navy Captain Charles Morgan, for questioning. "Very well, then," he asked, "who are you, and what were you about?"

De la Penne responded in English, "I am *Tenente di Vascello* Luigi de la Penne, Italian Royal Navy. And this is my second, *Sottotenente di Vascello* Emilio Bianchi, also of the Royal Navy."

Morgan asked again, this time addressing just the team leader. "Very well, Lieutenant, why were you mucking about under my ship? What was your mission?"

De la Penne responded, again only identifying himself and his second by name and rank, and then keeping silent.

Morgan tried again, this time addressing Bianchi. "And you, Lieutenant, why were you under my ship? What was your mission?"

Bianchi responded only with, "I am *Sottotenente di Vascello* Emilio Bianchi, Italian Royal Navy."

And so it went for the better part of two hours, the Italians telling Captain Morgan nothing.

It was already into the wee hours of the 19th, and Morgan decided that no British divers would be sent down to survey the vessel until daylight. He ordered the Italians confined to a compartment belowdecks; ironically, the compartment was just above the exact location where de la Penne had placed the mine.

Shortly before the mine was timed to detonate, de la Penne sent word to Morgan that an explosion was imminent, and that the ship should be evacuated. On hearing the report, Morgan was skeptical, but ordered the evacuation anyway. He then had de la Penne brought to him, and attempted to get him to reveal the mine's location. De la Penne still refused, and Morgan returned him to the same compartment (which Bianchi had never left).

The mine went off, and *Valiant* began to sink immediately. De la Penne was injured in the explosion, suffering lacerations to his arms and legs, while Bianchi went unscathed. Both men were brought up to the main deck in time to witness similar explosions under the *Queen Elizabeth, Jurvis,* and the *Sagona.*

Robby Cotton's ship—his first duty station—had just sunk into the mud beneath it. And Luigi de la Penne and his team had effectively disabled the British Mediterranean fleet for months.

Another sneaky attack from underwater, Robby thought, and felt ever more defenseless.

Gene Masters

6

Benghazi, December, 1941

Royal Italian Army *Sergente* (Sergeant) Marco Scarpetti never heard the explosion. The shell, which landed not forty feet from his position at the parapet, had blown him high into the air, and when he awoke, unharmed except for a massive headache and a buzzing in his ears, he was staring down the barrel of a British Enfield rifle.

"Hands up, you Dago bastard," yelled the soldier at the outer end of the rifle, "This war is over for you."

"So it would appear," Marco replied. The soldier was surprised to hear his Italian captive respond in English. It was 23 December, and just two days before Christmas, 1941.

In addition to his native Italian, the handsome, five-foot, six-inch, dark-eyed Marco was fluent in English, French, Spanish, and German, and he could read and write Latin and classical Greek. He had been studying languages at the University of Bologna for almost three years, when, in the Winter of 1935, Prime Minister Benito Mussolini persuaded Italian King Victor Emmanuel III to invade Ethiopia (also known at the time as Abyssinia).

The European states and the British had long since carved out vassal states in Africa, and Italy had been no exception, with colonies in Libya in North Africa, and Eritrea and Italian Somaliland on the horn of the continent. Between these last two possessions, and the only independent state left in Africa, was the 400,000-square-mile Ethiopia, whose head of state was Emperor Haile Selassie. Selassie claimed to be a direct descendent of the biblical King Solomon. Mussolini had in mind nothing less than the establishment of the new Roman Empire, and Ethiopia had to be conquered before it could be linked with Eritrea and Italian Somaliland to form the new Italian East Africa.

The British, who owned all the other surrounding colonies, had elected to stand apart from the fray in the hopes of keeping Italy in the British-French sphere of influence, and away from that of the Germans and Austrians.

When it became apparent to the Ethiopian Emperor that an Italian invasion was imminent, and that there would be no aid forthcoming from the British and the French, Haile Selassie issued the following general order:

All men and boys able to carry a spear go to Addis Ababa. Every married man will bring his wife to cook and wash for him. Every unmarried man will bring any unmarried woman he can find to cook and wash for him. Women with babies, the blind, and those too aged and infirm to carry a spear are excused. Anyone found at home after receiving this order will be hanged.

The Emperor was able to mobilize some half-million men. There were only about 400,000 rifles available for them, however, and many of these were of dubious age and in poor repair.

The Laconia *Incident*

* * * * *

For Marco Scarpetti, completing his third year of studies at the university, the siren call of *Il Duce* went completely unheeded. Dreams of military glory in North Africa and a new Roman Empire, together with the urgings of a certain Fascist professor for all his students to enlist, would never overcome Marco's overriding desire to finish his studies. In any case, in January 1936, to the dismay of his physician father and doting mother, their only son Marco was drafted into the *Regio Esercito,* the Italian Royal Army. His new superiors, noting his three years of university, had urged Scarpetti to apply for officer training, but Marco, an unwilling draftee in the first place, elected to serve as a private soldier.

The Ethiopian campaign served only to eliminate from Marco whatever illusions he might have remotely entertained about "a new Roman Empire." The poorly-armed and underequipped Ethiopians, with some soldiers fighting with little more than spears and machetes, managed to hold off a mechanized Italian Army and the Italian colonial troops for over a year. (In contrast, Germany would later defeat a well-equipped and well-trained French army in less than six weeks.)

It was then that the conquering Italians finally discovered that winning the war was one thing, while winning over the local populace was another thing altogether. Ethiopian Emperor Haile Selassie, forced into exile in Jerusalem, continued to hold sway over the Ethiopian people, even from afar. Rebellious guerrillas kept the *Regio Esercito* units continually looking over their shoulders.

If Marco was an unwilling draftee, he nonetheless made the best of a bad situation. He threw himself into becoming a good soldier with at least some of the same zest with which he had pursued his studies. He gradually rose through the ratings, and while still serving in Ethiopia, Marco made sergeant.

In May, 1939, Germany and Italy signed the "Pact of Steel," formalizing the Axis Alliance, and pledging military cooperation. Shortly afterward, there came the news of the coordinated German and Soviet invasion of Poland in September, 1939. It was that invasion that initiated war between Germany and Italy, and Poland's treaty allies, England and France. Soon afterward, Marco was transferred to serve in Libya with the Italian 1st Army.

For Marco, fighting an expanded war against other Europeans for God-only-knew-how-long, was the stuff of nightmares. Japan joined the Axis in September, 1940, with the Berlin signing of the Tripartite Pact with Germany and Italy, and Marco knew in his heart that it would only be a matter of time before the rest of the world would be drawn into the fighting.

Italy invaded Egypt from Libya almost coincidentally with the signing of the Tripartite Pact. Now Marco Scarpetti found himself fighting not Ethiopian spears and machetes, but British troops, and a trained and mechanized army. The following December, a British counterattack in the Libyan desert drove back the Italians, in the process decimating the

Italian 10th Army. Marco was lucky to have escaped the initial British onslaught, as he and the Italian 1st retreated north to Benghazi.

Now backed into a corner in Libya, Mussolini appealed to Hitler for assistance. *Der Fuehrer*, however, was much more concerned with the newly-initiated Russian campaign. Eventually, however, Hitler did respond to Mussolini, but only to placate his friend and Italian ally. He dispatched a small force to Tripoli under *Generallieutnant* Erwin Rommel in February, 1941.

In the months that followed, British, Italian, and German fortunes in North Africa ebbed and flowed across the Libyan and Egyptian deserts.

* * * * *

The British captured Marco just two days before Christmas. He was herded along with twelve other Italian POWs into a hastily-erected compound. Pointing to Marco, the British private who had captured him said to the officer in charge, "This one speaks English."

As their ranks quickly grew, Marco became the official translator for his fellow Italian prisoners, always the one selected to relay prisoner complaints to their British captors. The British, in turn, attempted to use Marco to relay their orders to the prisoners, something he steadfastly refused to do. The British soon labeled the diminutive, scholarly-looking, Italian as a somewhat-useful troublemaker.

What followed for Marco were five months of being shifted from prison camp to prison camp, as the fighting in North Africa raged on. Things were going badly for the British, who were being driven back into Egypt by the Germans, and the Italian POWs were only hindering the army's freedom of movement. So it was, then, that Marco was among some 400 other Italian prisoners transported to Alexandria. In Alexandria, the Italians were placed under a guard of Polish soldiers and then loaded aboard an aged Polish cargo ship.

After two days, the ship got underway, departing Alexandria for—*where*? The Italians had no idea where the ship was headed. Their Polish guards knew, perhaps, but none of the guards spoke any Italian. For once, Marco's language skills were useless—he knew no Polish whatsoever.

Conditions aboard the Polish ship were hardly ideal, with Marco and his compatriots confined to the ship's cargo holds. There were no bunks *per se*, but room enough to stretch out on the deck. The sanitary conditions were, to say the least, also primitive.

Nonetheless, the Italians were fed regular meals, and the fare, though simple, was reasonably nutritious. Each day, everyone was brought up on deck in small groups for two hours of sunlight and exercise. The Polish troops were not exactly gentle with their charges, but neither were they in any way brutal. Still, although many of the Italians became sick anyway, Marco managed to remain healthy.

7

Port Said, Navy House, January 1942

With three warships and an oiler sitting out of commission in Alexandria, the Royal Navy suddenly had about 1,700 excess sailors it needed to find billets for.

In Alexandria, Robby Cotton and his mates, Ralph Tinsdale, Charles Martin, and James Fellows, were given draft chits that posted them to shore duty in Port Said, Egypt, at the mouth of the Suez Canal. (Their mate, Arthur Kinkaid, was ashore, confined to the brig, having come down with a dose of gonorrhea. He had contracted the disease in Malta, but kept it hidden until there was nothing left to do but report himself into sick bay. He was rewarded with a three-week course of sulfonamide treatments and three months in the brig. To his dismay, Arthur later found out that if he had reported his ailment immediately, he would only have been confined to the ship until declared cured by the medics.)

For Robby and his mates, the 262-mile overland trip along the Egyptian Mediterranean coastline took six uncomfortable hours.

Upon reaching Port Said the day after New Year's Day, 1942, Robby, Ralph, Charlie, and James reported to Navy House, a large building that overlooked the wharf and harbor works.

Jim McLoughlin, Robby's other mate from *Valiant*, arrived several days afterward. There were no provisions for separating the ratings for sailors bunked at Navy House, and Robby was surprised and delighted when Jim McLoughlin threw his duffle down next to the empty bunk across from his.

"Hey, Minnow," McLoughlin said, "no telling what riff-raff they let in here, eh?"

"Jim!" was all Robby could think to say.

"In the flesh."

* * * * *

"I don't guess I'm much of a warrior, Jim," Robby confided to McLoughlin as they both walked guard duty along the pier at Port Said. "I was pretty much frightened out of my wits when the *Barham* was hit. First she was there, going along just fine-like, and a minute later she was all aflame and coming apart! And—God help me, Jim—all I could think about was that *Valaint* was gonna be next. I mean, I can handle me own in air raids all day long, 'cause at least we can shoot back at the bloody planes . . . but subs! There's no knowing, no telling . . ." (Robby paused for a breath, as McLoughlin listened.) "The men in the water," he continued, "dead bodies in life jackets floating—*Barham* just rolling over and disappearing like that—and all I could think about was, 'It's all over, I'm to be next!' " He paused again, and then said, the sadness palpable in his voice, "I'm just a bloomin' coward!"

"Bollocks, Minnow," McLoughlin reassured him, "you're no coward. You're just normal is all." McLoughlin paused for a few seconds, as if gathering his thoughts. "You'd be a lunatic not to have the shite scared out of you seeing a thing like that. You're normal, is all," he finally reiterated. "I was every bit as frightened as you was then, and I've seen a lot more of such then you. You don't ever get over being scared, you just get *used* to it. And you're always thankful that—for whatever reason—you're the one what's been spared. You just thank God for it, and move on."

"I sure hope you're right about that."

"I am, Robby. Trust me, Minnow, I am."

* * * * *

Robby, his mates, and McLoughlin would languish in Port Said for another five months, enduring what was a very hot and humid summer. With frequent liberty, they also got to know the city fairly well. Port Said offered many cultural venues to be seen in what was, after all, one of the more cosmopolitan cities in Egypt. Despite themselves, Robby and his friends couldn't help but admire and absorb at least some of the city's arts and architecture. They even learned a bit of Arabic. That is not to say that Robby wasn't inclined to give in to his baser urges.

He and his mates spent most of their liberty hours in the "recreational" parts of the city, partaking of its diversions. When Robby was at his weakest, McLoughlin was usually there to pound some sense into his head, warning him of the "billons of little bugs" that even the most alluring of the bar women carried, just waiting to infect him with all sorts of incurable diseases. As far as Robby's other mates were concerned, though, McLoughlin was just some sort of monk—or perhaps he was planning on becoming a priest after the war. "You're only young once, Mate, and there's a war on. You gotta pick them daisies when you can," was the advice that Ralph, Charles, and James proffered.

But McLoughlin wasn't always there. So, succumb Robby did, on occasion. He was never so drunk, nor so stupid, however, as to not wear protection.

The officers who gave the monthly VD lectures had always said, "Don't. But if you *do*, always, always, wear a condom. Piss right afterwards, and wash your peter down good when you get back to the ship." Robby knew that the Catholic Church said using condoms was a mortal sin, but fornication was also a mortal sin, and so, if he was going to go to Hell anyway . . .

8

Off Lorient, Vichy Republic, April, 1942

U-156 motored out to sea leaving the submarine pens at Lorient behind. It was 22 April, 1942, and newly promoted *Kapitanleutnant* Werner Hartenstein was taking his boat out on her third war patrol. Behind Hartenstein, painted on the steel cowling that enclosed the boat's conning tower, was his U-boat's emblem: a silhouette of the boat riding a lightning bolt.

Although it was April, and the sun was bright overhead, that Wednesday morning it was unseasonably cold. But Hartenstein, standing on the bridge of his command, was oblivious to the chill. He was in his element. He was taking his beloved boat, and a crew that idolized him, into the Caribbean Sea to do battle with the enemy.

Beside Hartenstein on the bridge was his first officer, *Oberleutnant zur see* Paul Just. Just had served as Hartenstein's first officer since the boat's commissioning, and Hartenstein had grown to trust him implicitly.

"It's a truly beautiful day, Captain," Just said, as he scanned the sky overhead with his binoculars, searching for their pathfinder escort. He could hear the aircraft's engine, but he could not spot the plane.

"Don't bother, Paul, you won't spot him. He's in the Sun," Hartenstein said, his boyish good looks somewhat distorted by his smile. (Whenever Hartenstein smiled, it seemed to somehow puff out his cheekbones, emphasizing the dueling scar on his left cheek and the sharpness of his chin. Whenever he smiled broadly, he looked, in a word, cadaverous.)

"Ah," Just exclaimed, "that explains it," and he set the Zeiss binoculars down to let them hang at his chest. *Oberleutnant* Just, at just under six feet, was half a head taller than Hartenstein, and a good deal heavier.

Lorient was on the Le Blavet River in West-central France. Once *U-156* cleared the estuary, Hartenstein turned his boat due west, on course to pass the island of Groix, well to the south. What had been a calm sea outside the harbor, was now choppy, with white foam caps atop ragged blue-violet waves blown by a brisk wind.

U-156 was a type IXC U-boat, a longer-range and more-capable version of the *Kriegsmarine's* workhorse submarine, the type VIIC. On that April morning, she was making top speed on the surface, just over eighteen knots, as fast as her twin, 4,300 bhp, MAN 9-clyinder diesels would drive her twin screws.

Before the morning was over, Hartenstein would exercise the boat and her crew, crash diving her to her maximum operating depth, 230 meters (750 feet). If he could, while submerged, he would live-fire a torpedo. However, the twenty-two G7e torpedoes *U-156* carried to her patrol grounds were far too precious, so, instead, he would exercise his torpedomen and the boat's six torpedo tubes (four forward and two aft) by firing water slugs from an empty tube.

The Laconia *Incident*

Once U-156 was back on the surface, her captain would have his crew exercise her armament: her 10.5 cm. deck gun, and her two (one 3.7 cm., and one 2 cm., twin-mount) anti-aircraft machine guns. But all that was for later in the morning. For the time being, Hartenstein would enjoy the crisp early morning air, the surging sea, and the sunshine.

Gene Masters

9

Ascension Island, June, 1942

Wideawake Field was ready for business, and Colonel Ross O. Baldwin, United States Army, an infantry officer, surveyed his new command. The construction that had begun on Ascension Island the previous April, under a cloak of tight security, was now complete. Baldwin looked out over the airfield from atop its air control tower, the highest man-made structure on the island. Of course, it was dwarfed by the 2,800-foot high peak of Green Mountain at the island's center, but Baldwin was of a mind that if the Army had been tasked to build that as well, it too would have been built in record time!

The 38th Engineer Combat Regiment, Baldwin thought, had done *"One hell of a job!"*

All the construction materials and equipment—every last bit of it—had to be off-loaded from deep water cargo ships at anchor, offshore, and onto smaller vessels, to land them, because Ascension Island had no harbor. Everything—heavy construction machinery, even the guns for the island's defense—had to be painstakingly ferried ashore.

Baldwin looked out from the tower over the new 6,000-foot runway. *Beautiful,* he thought, *just beautiful!* Besides the runway, he surveyed the camouflaged fuel storage tanks, the desalinization plant, gun emplacements, barracks, a hospital, and the other support facilities. None of these existed just ninety days earlier. There were also, he knew (though they were outside his field of vision), even two radar stations in place, one on either side of Green Mountain.

10

Port Said, June, 1942

Robby Cotton, his mates, Ralph, Charles, and James, along with Jim McLoughlin, were sure that the Royal Navy had completely forgotten about them. They had had been posted to Navy House since the beginning of the year, and it was now the end of June. The duty had been light: limbering the shore defense guns daily; cleaning the barracks incessantly; and walking guard duty around the base perimeter and on the docks.

Once again Robby and Jim had drawn guard duty on the pier. As they walked the long dock, they passed beneath a Polish ship that had come into port just the previous morning. The ship was old and rusty, and Robby was unable to read the name that was once brightly painted in white on the fantail. Now all he could make out was a faded initial "E" and most of an "A" about halfway through. He knew the ship was Polish, only because it flew the Polish flag: two horizontal stripes, white over red, the Polish royal crest in the center of the white stripe. There was that, of course, and then there was the contingent of Polish troops that had arrived with the ship and were immediate visitors to the ship's store at Navy House. They stood out. The color of their uniforms was much the same as the olive-drab of the British, but the cut was markedly different, and the high black boots they wore made them look quite distinct. The ship reportedly was transporting Italian prisoners of war from Libya, en route to South Africa.

Gene Masters

McLoughlin and Robby walked by a Polish soldier who sat on a bollard, smoking. "What ship?" McLoughlin asked, but received only a blank look in return from the soldier. He then repeated the question, louder and slower, as if that would somehow aid the man's English comprehension.

"Kominsky," the soldier said, pointing to himself. "Kominsky, Stanislaw," he added, and extended his hand.

McLoughlin shook it, pointing to himself as well, and said, "Jim," and then, pointing at Robby, said, "This here's Robby."

"Jim," Stanislaw parroted, smiling broadly and shaking McLoughlin's hand ever more vigorously, and then turned to Robby, saying, "Thisyearsrobby."

"No," Robby said. "Just Robby."

Stanislaw looked at McLoughlin quizzically, still shaking his hand, and said "Justrobby?"

McLoughlin couldn't help but chuckle. "Robby," he said.

"Ah. *Robby!*" Stanislaw released McLoughlin's hand, took Robby's, and repeated, "Robby!"

Then he said *"Miło mi cię poznać"* [Pleased to meet you in Polish]

Now it was Robby's turn to look confused. McLoughlin separated the two men, and, once again expounding the theory that anyone can understand English if you speak loud enough, shouted slowly and clearly to Stanislaw, "Guard duty. Need to walk pier," pointing up along the pathway he and Robby had been travelling. When the Pole only returned a blank look, McLaughlin shrugged, braced, and saluted him.

Stanislaw, in turn, braced, put on a game face, and returned the salute. McLoughlin then nudged Robby back along their guard route.

When their rounds took them back to the spot where they had met the Pole, he was gone. Neither McLoughlin nor Robby ever learned the name of the Polish ship—not even after they learned they were to soon board it and accompany the Poles and their Italian prisoners to Durban, South Africa.

Aboard the unnamed ship, locked away in an almost airless hold, deep in its bowels, was a diminutive young Italian Army sergeant named Marco Scarpetti. Along with Marco were some 400 of his compatriots.

*****

At about the same time the unnamed Polish ship was docked in Port Said, the British troop transport *HMT Laconia* was docking in Malta, not far from where *HMS Valiant* had once been berthed. Belying her once colorful, proud, and noble appearance, the *Laconia* now sported a dull gray paint scheme. The luxurious appointments that once coddled international travelers had been removed and replaced by more numerous and more practical accommodations. One of the more practical items now aboard the *Laconia* was a single, six-inch gun mounted on her stern. She had arrived there to embark British civilians and foreign service personnel for eventual transport back to England. Also being embarked were some British Army and RAF personnel who were bound for North African duty posts, or for rotation back home. Captain of the *Laconia* was Rudolph Sharp, late of the *HMT Lancastria*.

Headed back to England, and settling into a sparse interior cabin on the main deck, were RAF ambulance driver Donald Logan, his Maltese wife, Violet, and their newborn daughter, Helen. Donald was an RAF ambulance driver who hailed from the difficult to spell and impossible to pronounce town of Ynysybwl in Wales. His wife, Violet, was Maltese, and they had met and married in Malta.

The Logans had expected to leave port soon after they settled in, but were disappointed. Malta could become hot and uncomfortable in June, and while their cabin was comfortable enough, it would have become more so if the ship was moving, and there was at least a sea breeze.

When the ship eventually got underway from Malta three weeks later, Violet asked Donald, "Why ever are we headed east to the Suez Canal? Wouldn't it be quicker and easier to get to England if we traveled west past Gibraltar?"

"So it would seem, my love, so it would seem," Donald Logan agreed. "But I'm sure the Service has its reasons—probably something to do with Nazi subs in the area, or some such."

Violet looked unconvinced. And so, also, deep down, was Donald.

11

Lorient, Vichy Republic, July, 1942

Werner Hartenstein watched with great satisfaction as one of his junior officers, *Leutnant zur See* **Klaus Herbst** skillfully guided *U-156* into its slot in one of Lorient's submarine pens. He smiled at the thought that he was out of uniform, the epaulet on his jacket being a flat silver with two pips—that of a *Kapitanleutnant*. He had been promoted to *Korvettenkapitän* on the first of the month, and his epaulet should have been of silver braid.

U-156 had just completed her third war patrol, and Hartenstein and his crew had been wildly successful. The Caribbean had proven itself to be a fertile hunting ground. *U-156* had scored twelve solid kills, had damaged another cargo ship, and had disabled the American destroyer, *USS Blakeley,* having blown off her bow.

Waiting on the quay was a welcoming contingent from the *Befehlshaber der U-Boote,* or *BdU* (Commander of U-boats), flown in from Paris. Among the dignitaries was the Commander himself, *Vizeadmiral* (Vice Admiral) Karl Dönitz. Hartenstein and Dönitz were well acquainted; the admiral made it a point to get to know all of his U-boat commanders personally, and he genuinely liked this aristocratic gentleman from Saxony.

Once the gangway had been placed, the admiral was the first to board the boat. "An excellent patrol, Werner," Dönitz said, as he shook Hartenstein's hand.

"Thank you, Admiral," Hartenstein replied, "but with such a target-rich hunting ground, it would have been impossible not to do well."

"Be that as it may, Werner, you and your crew have nonetheless done the Fatherland proud."

Hartenstein could only respond with his cadaverous smile.

"And your home town has arranged a little celebration in honor of you and your entire crew, Werner."

Hartenstein's home town was Plauen, Germany, where he was born in 1908. Plauen, in Saxony, was in east-central Germany, near the Czechoslovakian border.

"A celebration, Admiral? Plauen has done this?"

"Yes, Hartenstein, Plauen! It seems that the town fathers have decided to fete one of their own with the *Fuehrer's* blessing. The Reich Chancellery has arranged for two private railcars to transport you and your entire crew across France and Germany to Plauen. While your boat is in refit, you will be riding in style."

"But there is much to be done, Admiral," Hartenstein protested. "The boat needs—"

"Relax, Werner. Paul [*Oberleutnant zur see* Paul Just] will see to all of it. He will be staying behind. Then Just will be reporting for U-boat commander training, taking command of *U-6* at the sub school. When he is ready, there will be a U-boat battle command for him."

Hartenstein smiled. "Paul is a good man, Admiral, and a first-rate U-boat officer. He will not disappoint." He was pleased for his first officer, but was disappointed to be losing him. Paul Just was a man he had grown to trust, a man he knew he could rely upon. Now he worried about the quality of his replacement.

"I agree." Dönitz paused, as if weighing his next words. "When you return from Plauen, *Oberleutnant zur see* Leopold Schumacher will become your new first watch officer. Understood?"

"Yes, Admiral. As you wish." Hartenstein knew very little about Schumacher, but the U-boat community was very small and very tight. He knew at least that Schumacher was from an aristocratic family, and that he had served with honor in U-boats out of Saint Nazaire.

"I have great plans for Schumacher as well, Werner, and I want him to learn from the best." Hartenstein received the compliment expressionless, and without comment.

Dönitz then made ready to leave the boat, and Hartenstein saluted: open right hand to the visor of his cap. The admiral returned his salute in kind. The open-palm, stiff-extended-right-arm, Nazi salute was used only rarely in the *Kriegsmarine*.

* * * * *

The 20 July affair at Plauen was such that Hartenstein was not sure it was all worth the long train ride. His parents had passed on, his siblings had long since vacated the place, and, as he was a confirmed bachelor, there was no wife nor family in Plauen to greet him. The day itself was dreary, the air hot and sticky, with an overcast sky and rain threatening, but, thankfully, never materializing.

There was a short parade through the town from the railway station, the officers and crew of *U-156* marching to the beat of an *oom-pah-pah* band, the street lined with cheering townspeople waving little Nazi flags: the black bent cross in a white circle, centered on a red banner. At City Hall, the squat, fat mayor gave a lengthy speech, praising their heroism and extolling their U-boat's exploits. The mayor then gave Hartenstein an oversized key to the city as the crowd applauded, and the city fathers looked on approvingly. When the speeches were over, there was plenty of beer and even some sausages for all.

Some of the submariners caught the eye of some of the more willing among the ladies of the town. And while Plauen was not all that big, there were many private and quiet spots available where the young ladies could do what they regarded as their patriotic duty.

When the crowd finally dispersed, Hartenstein and his men returned to the rail station, and boarded the train for the long ride back to Lorient and their boat.

Back in Lorient, *Oberleutnant zur see* Leopold Schumacher had already reported aboard for duty.

12

En Route to Durban, July - August, 1942

The passage through the Suez Canal, around the horn of Africa, and down the west coast of Africa to Durban, Union of South Africa, was long and tedious. The sea itself always seemed the same: listless, languid, green. The sea air was oppressive, salty-wet, and hot—even at night. The weather never broke until after they had rounded Mozambique, and headed into the Madagascar Straight. Even then, it was never exactly cool.

The unnamed Polish ship was old and slow, and her engines broke down continually. Robby swore they spent more time drifting in the open sea, or at anchor, while the crew worked on its ancient reciprocating steam engine, than actually sailing under propulsion.

Robby, McLoughlin, and their mates, Ralph Tinsdale, Charles Martin, and James Fellows, were nominally on board to man the single, aged, 4.5-inch gun mount affixed to the ship's stern. But the chief petty officer in charge of their little band spent most of the trip in his cups, and the gun was actually limbered no more than a dozen times during the entire trip south. There was, thankfully, never the requirement to actually aim it, let alone fire it, at anything threatening.

It seemed that only Robby fumed in anger at their situation. There they were, far from the fight, and not even training with the one weapon they had aboard. He had joined the Navy to fight—specifically to shoot

Germans—and now he and his mates were playing nursemaid to Poles and their Italian prisoners. He felt useless, his talents, such as they were, being wasted. "Relax," McLoughlin had advised, "we'll be in the thick of it again soon enough."

"We might at least shoot some with the bloody thing," Robby had countered, referring to the barely-used gun. "It's all we have to defend ourselves with."

"Defend against what?" McLoughlin asked. "There ain't a bloody enemy plane within miles. Any sub in the area would be lost. Better we get whatever rest and relaxation we can, while we can!"

Robby had no answer for that, but was still not mollified. And the weather was exactly what one would expect in equatorial East Africa at that time of year: hot—stifling hot. The daytime sky was filled with blazing sunlight, and the night sky with the wet, hot, fetid breath of Africa. Even the sea, green and oily, seemed to have been lulled into a languid stupor.

Robby and McLoughlin ran into the Polish sergeant they had met on the dock at Port Said on their second day out. Stanislaw recognized them immediately, and greetings and smiles were exchanged, if not in words mutually understood.

Over the next few weeks, the three men became ever more friendly, with McLoughlin, Robby, and Stanislaw introducing them to their respective mates. The British contingent learned some Polish, and the Poles some English.

Their Polish keepers had no particular beef against the Italian POWs. It was the Germans and the Russians who had raped and pillaged Poland, and it was the

Germans and the Russians that the Poles hated. The Russians were now nominally on the Allied side, but were still just as hated. The Italians, on the other hand, might have allied themselves with the Germans, but had never set foot in Poland. The Polish troops aboard the unnamed ship, therefore, actually sympathized with the plight of the Italian POWs.

The Poles brought their charges in groups of about forty to fifty up onto the main deck for sunlight and air for at least an hour or two, every day, and throughout each day. As far as Robby could see, the Italians were well fed, and looked healthy enough, all things considered.

The ship had just rounded the Horn when one of the prisoners, brought up from below for some sun, called out, "Hey, English!" to Robby and Jim, who were loitering on the man deck at the time. McLoughlin, noting that Stanislaw was among the Poles in apparent charge of the Italians, nodded to his friend, effectively asking permission to engage the Italian. Stanislaw nodded in return, and, Jim, with Robby in tow, walked over to the diminutive Italian who had called out to them, and extended his hand.

"Jim McLoughlin," he said, "and this here's Robby Cotton. The Pole is Stanislaw." Stanislaw stood back, wary of fraternizing with a prisoner.

"Marco Scarpetti," the man replied, shaking Jim's hand. "How do you do?"

"Fine, mate," Jim answered.

Then Marco grasped Robby's hand, which Robby had pointedly *not* proffered. "And how do *you* do?"

"I'm fine, mate," Robby said, leery of the man—an enemy after all. Hadn't he shot at Italian planes in Malta? He wondered if the man had picked up the greeting from an English phrase book, or actually spoke some English.

"Glad to hear it," Marco countered. "You look surprised that an Italian can speak English."

"I am—surprised, that is." Robby said.

"Me, too," McLoughlin chimed in.

Marco laughed. "Since one of your blokes captured me, I have had some opportunity to practice what I studied at school."

Robby, now warming to the Italian, smiled back. "Well, as we always say, 'It's an ill wind that blows no good.'"

"Yes," Marco said, "I have heard of that idiom."

Robby, not sure what an "idiom" was, said, "I'll bet you have."

"But that's the first I have heard anyone say it," Marco allowed.

Robby just beamed back, still smiling, not knowing what to think. There was, after all, an Italian whose English, he suspected, might be better than his own.

Stanislaw, all the while speaking effectively no English, could only stand by and observe the exchange. Jim, Robby, and Marco, each exchanging their stories, thus became acquainted.

Later, that evening, back below in the hold of the ship, Stanislaw approached Marco. "English?" he said. "You can teach?"

And Marco smiled.

The Laconia *Incident*

* * * * *

As the unnamed Polish ship made its way down along the East African coast, the *Laconia* was paralleling its route. It was moving faster, but made more, and usually longer, stops along the way, bringing aboard supplies in Aden, refueling in Mombasa, always discharging and picking up passengers. And at each stop along the route, in a seemingly never-ending stream, the ship's holds were being filled up with Italian prisoners of war.

Violet Logan grew more and more frustrated as the days wore on. She knew she, her husband, and their baby could have been in Wales weeks ago. *If only this bloody ship was headed in the right direction! And now the ship is loaded with Italians! The same people who are bombing my home in Malta!"*

Donald Logan counseled patience, but he, in truth, was even more frustrated than his wife. He had learned that the *Laconia*, now headed for Durban, South Africa, would languish there for several weeks before hitting yet another port: Cape Town. Only then would she venture out into the Atlantic and head for England. The ship had been comfortable enough, but he was longing to see his wife and child settled in their new home in Wales, and he was anxious about the upcoming passage through submarine-infested waters.

Gene Masters

13

Durban, August 1942

When the unnamed Polish ship reached Durban, on 15 August, Robby noted that the weather had finally gotten cooler, even in the full sun, for they were in the throes of what passed for winter in South Africa. He, McLoughlin, and their mates, Ralph Tinsdale, Charles Martin, and James Fellows, learned that the Italian prisoners, their Polish guards, and they themselves were to be transferred to a British troop ship due to arrive sometime in the next two weeks. Meanwhile, for Robby and his mates, their current quarters aboard the Polish ship would have to do.

Not so for the Italians and their Polish guards. They were marched off to a warehouse just off the pier. There, the Italians would join some 1,400 other (more or less) Italian prisoners of war who had been warehoused in Durban, all under the watchful eyes of the Royal Army. The POWs, now all of about 1,800 of them, were awaiting transport to England, destined be interned there as farm workers.

On liberty in Durban, Robby, McLoughlin, Tinsdale, Martin, and Fellows were enjoying some frosty South African Castle Lager beer, when who should walk in but the Polish guard, Stanislaw, and several of his mates.

"Stanislaw," Robby called out, "come join us!" The Poles, recognizing the English sailors who had cruised from Port Said with them, happily joined the five sailors.

"Hello, mates," Stanislaw greeted them. "Good to see you! But you are way before us with the beer. We are must be catching up."

"Stanislaw, my lad," McLoughlin opined, "your English is improving. I think I just understood what you said!"

"I learn English well, Jim." Stanislaw beamed. "Marco is teaching."

Robby pictured the little sergeant in the ragged Royal Italian Army uniform. From what little contact he and the others had had with—Scarpetti, was it?—he had, after all, found him to be a decent enough bloke—for an Italian. Too bad they were on opposite sides in this miserable war. Otherwise they might have become real friends. Then he thought better of it. Marco was obviously well educated, and even more obviously from the Italian upper crust, while Robby was just one of the common working men back home. Were it not for "this miserable war," they would never have met in the first place.

* * * * *

HMT Laconia, a battle-worn, old, gray lady, tied up alongside the pier at Durban, Union of South Africa, on 28 August. In her glory days, the *RMS Laconia* was a Cunard-White Star luxury passenger liner, plying the route between Liverpool and New York. At the onset of the war, she was requisitioned for the war effort by the British government, and converted for use as a troopship. This included removing all the signs and accouterments of luxury travel, and painting over the

White Star logo on the stack, and covering her sparkling white-with-blue, pin-striped colors with a dull merchant Navy gray. They had also mounted a discarded, Royal Navy, six-inch gun on the stern. The *Laconia* was a much bigger ship than the ship Robby and the others had ridden from Port Said. She was over 600 feet long, and displaced some 20,000 tons.

* * * * *

From dockside, looking up at their new home, McLoughlin exclaimed "I know that ship!"

"What?" Robby said.

"That ship—I know her. They've covered her up with that ugly gray paint, but I'd recognize her lines anywhere. She's the old *Laconia*. My dad was her chief steward, worked her on the route between Southampton and New York. I was all over her when I was a tyke!"

Once aboard, however, McLoughlin's thrill at seeing the old girl was muted. Gone were her garden lounges, potted plants, and gracious stairways with their polished teak railings. The old *Laconia* even had a lounge that was the authentic replica of a fully stocked English pub. Those were, instead, replaced with the austere requirements of a ship at war. Designed for no more than 2000 passengers, she could now be crowded with 2700. Robby noted that, in addition to the ship's full complement of lifeboats, there were also wooden rafts lashed to the deck; these would supplement the lifeboats in a genuine emergency.

Robby and his mates were quartered belowdecks in the stern of the ship, and assigned for duty under one

Royal Navy Lieutenant Percy Tillie. Their quarters had been fitted out in the newer style, with actual bunks to sleep in; hooks for their hammocks were replaced with canvas mats fixed to rectangles of steel pipe, topped with a thin mattress. Four high, these bunks could be folded back and trussed up out of the way during the day.

"What do we do with our hammocks?" Robby wondered aloud.

"Make a good pillow," McLoughlin volunteered.

There were other servicemen aboard — Army and RAF — but they were just passengers, of whom nothing was expected. And this was a ship, after all, and Robby and the others were Royal Navy, so, this would be a duty assignment. The *Laconia* was a DEMS (a Defence Equipment Merchant Ship), which merely meant she was armed. On her stern was that six-inch gun, bolted to the deck. And Robby, his mates, and the other navy men aboard, were assigned to maintain and man that gun under the watchful eye of Lieutenant Tillie.

From the dockside warehouse, the Italian prisoners and their Polish guards filed sullenly aboard *Laconia*. The Poles who had sailed with Robby and his mates from Port Said had been augmented by Polish officer cadets, requisitioned for that duty by the Royal Army brass that had overall responsibility for the POWs. The addition of the cadets brought the number of Poles aboard to 103. The Italians were to be housed under lock and key, deep in the bowels of the ship.

Donald Logan, his wife Violet, and their five-month-old baby daughter Helen, occupied an interior cabin on *Laconia's* O1 deck. Other British passengers on board, and headed home, included some diplomats, some civil servants, and their families.

*****

"Are we ready for sea?" Captain Sharp asked his third officer, Thomas Buckingham. Buckingham had assumed the duties of the second officer, who had been transferred to hospital ashore with a virulent case of Malaria. The day had dawned clear and sunny, the weather reasonably cool.

"Yes, Sir, Captain," Buckingham replied. "The ship is ready for sea. The Pilot is aboard, the gangway's up and secured, all shore water and power are secured, and all mooring lines are singled. The first officer reports the propulsion plant is on line and ready for sea."

"So, we are ready for sea, are we?" Sharp commented, addressing nobody in particular, and gazing up at the sooty black smoke spewing from the ship's single stack. Buckingham was equally dismayed at the sight, but knew that the first officer, who was the ship's chief engineer, had done all he could to clear the stack emissions. But the old ship was at least a year and a half overdue for engine overhaul, and the fuel oil that had been loaded aboard in Mombasa had been of dubious quality.

Three days later, the ship docked in Cape Town, where it took aboard additional passengers, mostly civilians: bureaucrats and their families, some Embassy staff. Cape Town was further south than Durban, almost as far south as one could travel and still be in

Africa, but it was, in fact, a hundred miles west of L'Aqulhas, Africa's southernmost city, set at the very southern tip of the continent. The weather there was only slightly cooler than it had been in Durban, but the days were just as clear and sunny.

The *Laconia* loitered in Cape Town for just two days, but Robby, McLoughlin, and the others got to pull liberty there, finding the beer much as it had been in Durban.

They were just getting the lay of the land, when it was time to leave; this time the destination was Liverpool. Jim McLoughlin was overjoyed with the prospect of returning to his home, and told Robby as much. It was 4 September.

Aside from Laconia's 463 officers and crew (which included Robby's lot), aboard were 87 civilians (including women and children), 286 British Army and RAF personnel, 1,793 Italian POWs, and their 103 Polish guards, Stanislaw Kominsky among them.

And among the Italians held below, deep in *Laconia's* third hold, was Marco Scarpetti.

14

Mideast Atlantic, August, 1942

U-156 had departed Lorient for its fourth war patrol on 20 August. At sea, conditions were much as they had been when the boat had departed on her last patrol: the sea slightly choppy, the weather still clear, but now much warmer. Vice Admiral Dönitz had told Hartenstein that he was assigning his boat for this patrol to some of the quieter waters of the Middle Atlantic; both he and his crew deserved "a little rest."

Hartenstein and the crew of *U-156* had thus been ordered to patrol the waters south of the Bay of Biscay, and off the west African coast. The area was not deemed to be target rich, unlike the Central Atlantic waters assigned to *U-156* on her third war patrol. But Hartenstein and his crew had already drawn blood. Just one week out of Lorient, *U-156* had sunk the 6,000-ton *SS Clan Macwirther* just off the coast of Casablanca. The weather had stayed pretty much as it had been off Lorient, hot and clear, with the sea, again calm, with gentle swells and a slight chop.

As with *Laconia*, the *Clan Macwirther* was a DEMS ship, and had been carrying manganese ore, linseed, pig iron, and assorted general cargo. Two torpedoes, fired from a submerged *U-156*, had sent her to the bottom in just ten minutes. Three lifeboats, containing 79 survivors, managed to escape the ship before she went down.

The *Befehlshaber der U-Boote*—the BdU—or Commander of U-boats, had issued standing orders that U-boat commanders should make every effort to locate and take prisoner the first and second officer of any ship sunk.

Then, following BdU's standing orders, Hartenstein surfaced *U-156* just as the *Clan Macwirther* slipped below the surface. He questioned the survivors in each of the lifeboats, pulling his boat alongside and calling down from the bridge. "My apologies for sinking your ship," he began, in his excellent English. "An unfortunate outcome of the war, you see. Is the captain or chief engineer aboard?"

"No," came the universal response. "They were aboard the ship when the she went down."

Hartenstein then asked the name of the ship, and the nature her cargo. After learning these details, he ordered that the survivors be given containers of water and some rations, and then personally gave them the course to the nearest landfall. (Those fortunate survivors would be later rescued by a passing British freighter.)

On departing the scene on the surface, the new first watch officer of *U-156*, *Oberleutnant zur see* Leopold Schumacher, said to Hartenstein, "I'm not at all sure that *Kreigsmarine* command would be sympathetic with your kindness to the enemy, Captain."

From the moment he reported aboard, Hartenstein had been taking the measure of Schumacher. The man was actually quite handsome, he had observed, with even features and startling brown eyes. Like Hartenstein, Schumacher had served in a torpedo boat in the Spanish Civil War. But Hartenstein had left for U-boat training before Schumacher began his service in Spanish waters, and so they had never actually met in that theater. What Hartenstein had seen of Schumacher thus far, however, he had approved. Still, here was yet another opportunity to feel the man out.

"Really, Leo?" Hartenstein replied warily. "But I am sure that doing my duty, attacking and destroying the enemy's ships, does not require me to abandon my humanity. Those British were men, after all, human beings just like us, just doing their duty. We must respect them for that, should we not? And show them whatever compassion we might hope to receive from them, were our roles reversed? Surely you can agree to that, can you not, Leo?"

Schumacher looked doubtful at first, but then, after a pause, finally said, "I can, Captain, and I do. I'm just not at all sure the admiral would agree. And certainly *Der Fuehrer* would not."

Hartenstein laughed outright. "I think, Leo, that you are selling Admiral Dönitz short! We are *Kreigsmarine*, Leo, after all. Mariners. And all mariners follow the code of the sea. We serve Germany, of course, and so does the admiral. We willingly fight for the Fatherland, just as Dönitz does, and we fight to win. But we also fight with honor. Regimes come and go, and this one may well last for a thousand years as *Der Fuhrer* promises, but we fight not for a regime, not for National Socialism, but for Germany! For the Fatherland! No, Leo, we who serve in the U-boats answer to the admiral. Let *him* worry about *Der Fuehrer*."

Schumacher again looked lost in thought. Now, Hartenstein knew, he would find out if he had read his first officer correctly. Finally, Schumacher said in response, "Food for thought, Captain. But you are certainly correct in that we serve the Fatherland, that we fight for Germany!"

Hartenstein smiled. "Very well, Leo. Then what say you, that for the time being, we just continue to do our duty, eh? We shall behave with honor, and let Admiral Dönitz worry about Herr Hitler!"

Schumacher nodded his assent, but Hartenstein sensed his first officer was still not in complete agreement. *Very well, Leo, I'll take that for now,* he thought.

With that, Hartenstein ordered *U-156* on a course south, in search of fresh prey.

15

Southeast Atlantic, September, 1942

The *Laconia* had departed Cape Town on 4 September, and Royal Navy Lieutenant Percy Tillie had, on their first full day at sea, decided to assemble his gun crews and exercise the ship's gun. The weather was much the same as it had been in Cape Town, just warmer, but still sunny and clear. The sea was calm, a sheet of bottle-blue glass.

It appeared that only Robby and Jim McLoughlin thought Lt. Tillie's attention to duty was a good idea. "Bloody hell," was all James Fellows could say. "What a bloody waste of time!"

"Come on, mate," McLoughlin said, trying to sooth Fellows and the others. "What else have we got to do?"

"No," James replied. "It's all of a cock-up. Tillie's having us fire shells at nothing whatever—there's no target out there. I think he's just trying to prove to hisself that this bloody wreck of a gun fires at all. All we are is a source of entertainment for the passengers—gives 'em something to gawk at to pass the time."

McLoughlin drew a deep breath as Fellows and the others rattled on.

"And a lot of good that popgun will do in a real fight, then, eh?" Ralph Tinsdale opined. "Here we are, in the middle of the bloody ocean. Bad enough we're blowing enough black smoke up in the air that we can be seen for miles by any enemy out there, and so we're poppin' off shells besides. Making enough noise so's the enemy can hear us, lest he can't see us! I tell you, Jim, if we don't get torpedoed, it won't be for our lack of trying. Nothin' but a bloody target, we are. Bullocks!"

"All the more reason to be practicin' at defendin' ourselves, then, eh?" Robby joined in, reasoning with his friends.

"Defendin' ourselves, are we?" Charles Martin laughed. "Not with that shite-slingin' piece of shite. Takes forever to train her 'round, and she's crewed by us pack of clowns, who've never even aimed her at a real target. Not proper guns like the ones we had on *Victory*! Like James here said, 'A bloody waste of time!' "

And to that, McLoughlin and Robby had no reasonable response that might calm their friends, possibly because, deep down, they agreed with them in part. The gun really was probably useless in a real fight. And for Robby, the thought of announcing their presence to lurking German U-boats was frightening. Once again, he saw the *Barham* inside his head. *The* Barham *had "proper guns" like the ones we had on* Victory. *Bloody little good it did them! If it wasn't Germans in airplanes trying their best to kill them, it was bloody Germans on U-boats!*

* * * * *

On that same first day at sea out of Cape Town, *Laconia's* passengers were issued life jackets and were assigned lifeboat stations. Lifeboat drills followed the next morning, shortly after breakfast. No lifeboat stations were assigned to the Italian prisoners in the ship's holds, nor were they issued life jackets.

The *Laconia* carried twenty-eight, thirty-foot-long lifeboats, mounted in seven double tiers on each side. They were sufficient to accommodate the ship's pre-conversion complement of passengers and crew; to accommodate the additional post-conversion human cargo, the Admiralty had directed that several wooden rafts be mounted on racks on either side of the ship. They were to be released and allowed to slide down the side of the ship, and into the water, in an emergency. Unfortunately, in order to board one of them, the survivors had to first be in the water themselves.

Now, with the Italians aboard, the *Laconia* carried well in excess of the number of people that could be safely accommodated were there to be an order to abandon ship. Worse, the pilfering of emergency rations and fresh water stored aboard the lifeboats was commonplace, especially by service people allowed on board in foreign ports. And the *Laconia* had visited several such ports en route to Cape Town.

*****

Third Officer Tom Buckingham took his newly assigned responsibilities seriously. Now third in line of command, after the captain and the chief engineer, he was expected to tour the ship on a regular basis and see to its proper running. Captain Sharp, for whatever reason, kept himself to his quarters, venturing out only on the bridge from time to time to take the air and curse the plume of black smoke spilling from *Laconia's* stack.

And, as for the chief engineer, he was rarely seen on the upper decks, as he was more or less stuck in the engineering spaces. He had his hands full with the aging and cantankerous propulsion plant, the very one producing that telltale plume.

For Buckingham, the upper deck passengers appeared to be faring well enough, so it was the Italian POWs that most concerned him. The contingent of British Army men aboard were technically responsible for the POWs, but neither the officers nor their men seemed to be the least bit concerned about the Italians' welfare.

Sanitary conditions for the prisoners down in the holds were, to say the least, primitive. There was little opportunity to exercise the prisoners, and little space to do it above decks. Showers consisted of bringing the men topside in small groups and hosing them down with seawater from a fire hose. Rations for the prisoners aboard *Laconia* were short, with several POWs on bread and water—punishment for some now long-forgotten infraction. Some of the men were sick, and needed medical attention, and there wasn't even deck space enough in the holds for the sick to lie down. Their Polish guards carried rifles, but had not been issued any ammunition.

Despite the language barrier, they appeared to Buckingham to be considerate enough of their prisoners, while still remaining stern enough to maintain good order. There was little that they could do for their charges, however, beyond seeing to it that they were fed their short rations on time, twice a day.

The Laconia *Incident*

And that same language barrier made Buckingham doing anything to alleviate the prisoner's suffering all the more difficult. He spoke no Italian or Polish, and worse, the Poles who were guarding the Italians, spoke no Italian whatever, and precious little English.

It was on the second day out that Buckingham discovered Marco Scarpetti, who was being held in number three hold. He had overheard the young Italian sergeant teaching English to one of the Poles, and, interrupting the lesson, asked, "So you *can* speak English?"

"Yes Sir," Marco replied, his English only slightly accented, "I do."

"Excellent," Buckingham said. He eyed the short, dark-eyed Italian in his soiled, tattered uniform, the three stripes on his arm patch barely distinguishable. "And what is your name, Sergeant?"

"Marco Scarpetti, Sir, First Royal Italian Army."

Pleased, Buckingham turned and addressed Scarpetti's student, a Polish sergeant who appeared to be somewhat older than his instructor, and asked, "And you Sergeant, can you speak English as well?"

"I speak little," Stanislaw answered, his English heavily accented. "I learn what Marco teaches."

"And your name, Sergeant?"

"Kominsky, Stanislaw, Lieutenant," he replied. "Polish Second Corps."

The "Polish Second Corps," Buckingham knew, was a corps of Free Polish Army troops who had been assigned to the British Eighth Army. Somehow, some of these men had ended up aboard *Laconia* and had more or less taken charge of the Polish army cadets who had been pressed into service in Durban to guard the Italian prisoners.

"I am pleased to meet you both, Marco and Stanislaw, you can be of great help to me, if you are willing. I speak neither Italian nor Polish, and you can help me communicate with both your fellow prisoners, Marco, and you, Stanislaw, with your countrymen, their guards."

Both men beamed, nodding an enthusiastic assent.

16

Ascension Island, September, 1942

Captain Robert C. Richardson III, United States Military Academy, Class of 1939, had, immediately upon graduation, requested assignment to the Army Air Force. He graduated from flight school a year later, and was so skilled a pilot that he was immediately assigned as a flight instructor at Randolph Field in Texas. It was at Randolph Field that he qualified in various training aircraft. Richardson moved on to a series of other assignments during his early career, including squadron command positions, and qualified in a variety of aircraft, including advanced twin-engine bombers, and fighters, including the P-40 Warhawk.

In April, 1942, Richardson took command of the 1st Composite Air Squadron, then headquartered in Key Field, Mississippi. The 1st Composite consisted of a flight of eighteen P-39 Bell Airacobra fighters, and five B-25 Boeing Mitchell bombers (the latter the same aircraft that Colonel Jimmy Doolittle flew off the *USS Hornet* to bomb Tokyo that same month).

The 1st Composite Air Squadron was transferred to Wideawake Field, on Ascension Island, in August, 1942, and was declared operational a month later. As commander of the 1st Composite squadron, Richardson was tasked with defense of the island and anti-submarine warfare patrols.

Colonel Ross O. Baldwin was still in command of the Army garrison on Ascension. Reporting to him, and officer-in-charge of the airfield, was Colonel James A. Ronin, and, in turn, Squadron Commander Richardson also reported to Ronin.

17

Southeast Atlantic, 10-11 September, 1942

Third officer Buckingham did what he could to improve conditions for the Italian POWs. Working through Stanislaw and Marco, he arranged for the sickest prisoners to be transferred to the few open beds in the ship's sick bay. There, if nothing else, they would at least get a decent night's rest.

With the cooperation of their Royal Army overseers, Buckingham arranged the cancellation of the bread-and-water punishment rations, and had the Poles ensure that the toilets be cleaned regularly—a provision for basic sanitation that, for whatever reason, had been entirely overlooked. He then set about providing crude showering and washing facilities, sparing the Italians the ignominy of their topside fire hose showers, and giving them the means for washing their clothes and bed linens.

For their part, the Italians took note of, and greatly appreciated, Buckingham's efforts.

On 10 September, the Admiralty ordered Captain Sharp to alter course the following day to a more westerly heading, away from the African coast, presumably to avoid possible enemy submarine activity. Ironically the new course put the *Laconia* miles closer to the track of Hartenstein's *U-156* now heading south in the general direction of Cape Town.

*****

In the early morning hours of 11 September, a flight of four-engine American B-24D bombers, the 343rd Bomber Squadron, passed Ascension Island en route from Brazil to the North African theater. The weather during their entire flight had been excellent, with moderate cloud cover, and unlimited visibility.

One of their number, piloted by Army Air Force Lieutenant James Hardin, developed problems with a smoking engine, and reversed course, landing at Wideawake field. For Hardin and his crew, this was the beginning of a course of fateful events.

*****

In the early afternoon of 11 September, *U-156* was cruising on the surface, heading south-southeast, when the starboard lookout reported a column of thick black smoke off the starboard bow. It was clear and sunny, visibility was excellent, and the sea was calm, with gentle swells.

"Where away, Willi?" Hartenstein asked the officer of the deck, as he climbed the ladder up to the bridge.

"There, Captain," Wilhelm Klempt, Hartenstein's fourth officer, said, pointing. "A point off the starboard bow."

"Nothing like a possible target advertising its presence," Hartenstein quipped, "Eh, Willi?"

"If it is a target, Captain," Klempt allowed. "It could well be a hospital ship, or some such, or worse, maybe a

The Laconia *Incident*

Q-ship, announcing its presence like that." Klempt was worried that the contact was possibly a raider, an innocent-looking ship that was actually armed to the teeth, luring an attacking submarine to its sure destruction.

"Let's just see, shall we?" Hartenstein said. "Bring the boat around to course three four zero, and speed up to 'full ahead both,' " thereby ordering a course and speed to close the contact.

"Aye, Captain, course three four zero, full ahead both," Klempt acknowledged the order. Then, calling down to the helm, "Rudder full right, new course three four zero. Increase speed to full ahead both."

A half hour later, as *U-156* closed the prospective target, the sound of the firing of the ship's gun reached the boat. "She's armed, Captain," Klempt observed.

"So she is, Willie, so she is. Well, if she has teeth, she's no hospital ship, and no Q-ship would so advertise her ability to strike. So we'll just watch her for a bit. But I am not anxious to come under her gun, so we had best not advertise *our* presence. Dive the boat, Willi," he said, and disappeared down the bridge hatch.

"ALARM!" Klempt shouted down through the open hatch, as the lookouts scrambled below. He quickly followed them below, securing the hatch behind him. Throughout the boat, as the alarm was passed, men performed automatically, their training superb. The cook in the galley grabbed the bar that operated the vent valve above him, swinging the bar and opening the valve, and then returning the bar to its neutral position. Elsewhere in the boat, the other vent valves were similarly opened by other sailors performing their assigned duty, opening the valves that allowed water to flood into the buoyancy tanks.

The boat quickly became heavy enough to submerge. In the control room, the diving planes were set to full dive. All off-duty crewmen made their way forward to the forward torpedo room, their combined weight making the bow heavy. In the engine room, the enginemen, responding to the shifted indicator on the engine order telegraph, secured the diesels and shut the engine intake valves by hand. And, in the motor room, the motormen shifted propulsion to the batteries.

In under a minute, *U-156* was under water, and Hartenstein was manning the periscope, observing, as the boat approached the *Laconia*.

18

Southeast Atlantic, 11 September, 1942

Aboard *Laconia*, on orders from Lieutenant Tillie, Robby, McLoughlin, and the others secured from gunnery drill, their mates still complaining about that gun's only reason for being was to alert the enemy to their presence.

"Bloody hell! Tillie might just as well set off fireworks from the fantail," came James Fellow's now-familiar rant, "or install bloody loudspeakers on our smokestack and play the bloody *horst wessel* song for the bloody Huns, and then announce 'Here we are, come and bloody get us!' "

This time Jim McLoughlin didn't have much to say to counter his friend's complaint, perhaps because he had come to realize he actually had a point. For his part, Robby would have been the first to admit he didn't know a lot about submarines, but he knew enough to realize that a torpedo could be headed for the *Laconia* at that very moment, and he wouldn't know it until it struck, and blew the ship out of the water. And, again, the very thought terrified him.

Once more Robby remembered the *Barham*; he had seen what a torpedo could do. And he didn't want to even think about it. *HMS Barham* was, after all, a "right proper" British warship, and, sure, it took three torpedoes to sink her, but the *Laconia* was as vulnerable as a rowboat compared to the heavily-armored *Barham*.

And the U-boat that sunk the battleship had gotten away clean, no matter that *Barham* was in the middle of a British task force.

So how much of a chance would poor Laconia *have,* Robby thought, *here in the middle of the bloody ocean, with not another friendly soul around? By the time we brought the gun to bear — and if we're lucky maybe even shoot her out of the water — what then? Old* Laconia *would still be a goner. Even if a body were to survive the attack, abandon ship, and make it onto a lifeboat, what then? The nearest land's got to be hundreds of miles away. And that's if you're even lucky enough to even get aboard a lifeboat! More likely you go straight into the water, and if you manage to do that and stay in one piece, who would be there to fish you out? Nobody. Best you were blown away in the first place — at least that would be quick — better than floating about until you just gave up, and drowned, or worse, became some shark's dinner. And, probably, just like with the* Barham, *the bloody U-boat would never get detected in the first place, and just slink away, not a care in the world . . .*

"Well, Minnow," McLoughlin finally said, breaking Robby's reverie, "the others must know Tillie's only doing his bloody job."

"Perhaps," Robby replied, "he is that, but you've got to see their point as well, Jim. It doesn't make a lot of sense to be making all that noise. And then there's that bloody trail of smoke, to boot. We're sitting ducks out here, in the middle of nowhere, and we'd be wise not to advertise our presence to whomever."

"To the Jerries, you mean," McLoughlin offered. "True, that, but face it, Minnow, we're just cogs in the wheel, along for the ride. But pity them poor Italians. If we do get hit, none of them will stand a chance. At least the two of us might make it into a lifeboat."

"If we're lucky," Robby answered. *Then again, maybe not*, he thought. *Dyin' in a lifeboat is still dyin'.*

* * * * *

Now that the noise of the gun being fired was over, and there was nothing to disturb the baby, Donald and Violet Logan were taking a turn on deck, Violet holding little Helen in her arms.

"Lovely day," Violet said. "I still can't get over the fact that here we are in the middle of September, and it's so sunny and lovely and warm."

"Don't get used to it," Donald chortled. "I can guarantee it won't be like this in Wales!"

"I guess not. I'll never understand why you British put up with such a miserable climate. Malta is just beautiful this time of year."

"You're right, of course," Donald readily agreed. "Perhaps that's why we left England behind so often and went off to establish an empire—that's the theory, at least."

"That and your miserable English food," Violet merrily retorted.

Donald chuckled. They had had this repartee before, many times. Now came the part where he said, "Yes, my dear, but all the more reason for us to marry good cooks..."

He barely got the words out when little Helen started wailing. "What's all that about?" Donald asked, anxiously.

"She's just hungry," Violet replied, "and probably wet as well. Stay and enjoy the stroll, my dearest, and your daughter and I will retire to the cabin for a little refreshment and a fresh nappy."

As she turned to leave, she was happy to see that her husband turned with her, preferring, rather, to stay with his little family and leave the pleasant weather on deck behind.

* * * * *

"She's British and she's armed, Captain," Leopold Schumacher said, as he peered through the periscope at the *Laconia*, "a proper target."

"Very well, then," Hartenstein replied. "We will wait until she's out of sight, surface, and then run ahead and cross her track for an attack from the east. We will strike just after sunset. That way, she will still be clearly visible in the western twilight, and we will be on her dark side."

"Yes, Captain," Schumacher acknowledged, smiling, pleased with his commander's battle plan.

Much later, at 7:54 PM, *U-156* was finally in almost perfect position for her assault on the British transport. The boat was on the surface, to the east of her target, the liner framed against a cloudless evening-twilight sky, just as Hartenstein had planned. The sea swelled gently, it's daytime, rich blue now shifting to liquid slate.

"Leo, Hartenstein ordered, "have the forward torpedo room prepare to fire three torpedoes—tubes one, two, and three."

The Laconia Incident

"*Jawhol, Kapitän,*" Schumacher replied, and relayed his commander's orders to the torpedomen forward.

At 8:04 PM, with *U-156* still on the surface, and now in perfect firing position, Hartenstein launched the first of the two torpedoes that eventually struck the *Laconia*. After sending a second torpedo on its way, Hartenstein decided to hold the third torpedo in reserve, and immediately ordered *U-156* to submerge. *Laconia* was, after all, armed, and if, for any reason the attack went awry, he was unwilling to expose his boat to the possibility of coming under her gun.

The first torpedo struck the *Laconia* at 8:07; the second struck just thirty seconds later.

Gene Masters

Part II

No Good Deed Goes Unpunished

1

South Atlantic, 12 September, 1942

"**He knew me dad,**" **Jim McLoughlin had said earlier,** prior to leaving Robby Cotton in their quarters, belowdecks in *Laconia's* stern, on his way to meet the ship's chief steward, who had promised him a gourmet meal.

Off duty, Robby was relaxing in his bunk, his rolled-up hammock serving as a pillow, and was looking forward to his own dinner in the crew's mess. Nearby, his mates, Ralph Tinsdale, Charles Martin, and some of the other off-duty sailors were also relaxing. James Fellows was in the head.

The food aboard Laconia *is really pretty good,*" Robby mused silently. *Can't see what Jim is so excited about...*

His reverie was rudely interrupted by the explosion forward. Robby had heard that sound before, and another picture of the *Barham* flashed in his brain. "Torpedo!" he shouted aloud, "We've been hit!"

The other men in the compartment, fear and surprise registering on their faces, reacted quickly to Robby's shout, and swung out of their bunks. A dazed James Fellows emerged from the head, hastily pulling up his trousers.

"Let's get topside and man the gun," someone shouted, just as the deck was rolling away beneath them, knocking Fellows off his feet and the others back into their bunks, or sprawling on the deck.

"Grab your life jackets!" someone else thought to shout, as the men picked themselves up, only to be knocked over again as the ship rolled back in the other direction, this time to starboard.

Robby was scrambling to hold his footing and pull on his life jacket, when the second torpedo hit. This time, the ship lurched in the other direction, settling at a port list far more severe than the first. He followed the others up the ladder, just forward of their compartment, as the sailors slowly made their way up and then back to the fantail.

When they reached the gun, they found Lieutenant Tillie already there, and the disgusted look on his face as he surveyed the scene said it all. The ship was now pitched over about twenty degrees to port, and since the gun was manually trained, there was no way it could now be trained around to aim it. That was, of course, assuming if they could find a target to begin with, and that the sub that had torpedoed them was kind enough to be on the surface somewhere within the scope and range of the gun. Nonetheless, Ralph and James had opened the ready locker and were doing their best — given the canted deck — to carry two shells over to load into the gun.

"Don't waste your time, men," Tillie called to them. "You're on a bloody fool's errand. If we do manage to load the bugger, how in blazes will we aim it? And at what?"

Just then, they heard the word passed, "Abandon Ship! Abandon Ship! All passengers will proceed in good order to their assigned lifeboat stations. Abandon Ship!"

The *Laconia* then lurched suddenly again to port, throwing Ralph and James off their feet, each man dropping the shell he was carrying. Tillie grimaced, Robby and the others cringed, all waiting to be blown to hell, as the shells hit the teak deck and rolled off to port. One careened off a railing column and went over the side and into the sea. The second lodged, stuck, in the scupper. Everyone let out their breath.

Tillie looked like he was ready to chew out Ralph and James, but then apparently thought the better of it, as Ralph made his way to where the shell had lodged, picked it up, and dropped it over the side. Tillie then said, "Right. Now lads, see to yourselves."

All then made their way forward, to where the ship's crew would hopefully be launching lifeboats. It was the last time that Robby would see Tillie, who set about helping the ship's crew launch the lifeboats. Fellows, he saw, had bullied his way aboard one of the boats along with some of the ship's passengers. Robby, along with Tinsdale and Martin, continued forward, but he and Tinsdale became separated from Martin somehow, and Robby and Ralph Tinsdale found themselves alone in the midst of a frightened and panicked crowd.

* * * * *

Tom Buckingham had the bridge watch, having just relieved the watch officer not twenty minutes earlier. Rudolph Sharp appeared on the bridge almost immediately after the second torpedo struck the ship and said to him, "There's nothing left to do here, Mr.

Buckingham. See to passing the word to abandon ship. Then you best go and attend to the passengers."

"Yes, Captain," Buckingham said. After passing the word to abandon ship, he left Captain Sharp behind on the bridge. He made his way as best he could, down askew ladders and a canted deck, heading to the main deck, there to see to the launching of the lifeboats.

Buckingham saw that launching the starboard lifeboats was difficult at best, although the crew managed to launch some of the boats on that side. Those boats, overloaded with half again as many passengers as they were designed to hold, were swung out as far as possible from the ship. Still, as they were lowered, the boats hit the angled side of the ship, and, skidding down the side, spilled out most of their passengers.

Of those who managed to remain in the boats, many were still spilled out, as the boats finally went into the swelling waters when their falls were released. On one of the starboard boats, the forward falls parted, and the boat swung away only to hang vertically, stern down, off the remaining fall. Its passengers were unceremoniously dumped onto the side of the ship, and then into the heaving water below. Most of those who eventually climbed aboard the boat were seriously hurt, and, of these, many were bleeding from cuts and open wounds.

Then, Buckingham made his way forward to the wooden rafts mounted there. He ordered them cut loose and sent down the ship's side and into the water. He hoped that none of them struck the people already in the water, but, to starboard, there was no way to ensure that the water below was clear before launching the rafts. So there was nothing else he could do. Buckingham hoped for the best, and then made his way to the port side.

* * * * *

On *Laconia's* bridge, Rudolph Sharp watched in dismay as the second ship under his command fell prey to the enemy, sinking beneath his feet. He had done all he could. No doubt, his first officer, the ship's chief engineer, was doing his best as well, striving to save the engine plant. But Sharp knew the man fought a losing fight.

Buckingham, good lad that, he will see to the passengers — perhaps even save himself, Sharp thought, and then turned and went into his cabin, locking the door behind him.

* * * * *

Belowdecks, the torpedo strikes sent the Italian prisoners and their Polish guards into a panic. In number three hold, inside their cages, the Italians surged toward the padlocked gates that penned them in. The very force of the bodies against the cage doors burst them open, but those in the forefront, not initially squashed to death against the doors in the panic, were subsequently either trampled, or crushed to death, by the surge of prisoners pressing from behind.

As the surviving Italians made their way to the passageway to the upper decks, their Polish guards — no ammunition for their rifles — fixed bayonets and held them back as best they could. The Poles were quickly reinforced by the British Army contingent, who fired on the POWs, desperately trying to hold them at bay — anything to keep the Italians away from the lifeboats until all the regular passengers were away.

Stanislaw Kominsky was among the retreating Poles. He hung back from his fellow guards as best he could, not wanting to appear unwilling to do his duty, but also not willing to kill any Italians — if he could possibly avoid it. He worried about his friend and teacher, Marco Scarpetti, but Stanislaw did not see him, searching for him as best he could among the advancing POWs.

* * * * *

The POWs and their guards in number two hold, where the first torpedo struck, were even less fortunate. Of those not immediately killed by the blast, many were drowned, as the hold filled with water when the ship rocked to starboard. The caged prisoners never did manage to escape their prisons; those not drowned outright were drowned when the *Laconia* finally sunk.

* * * * *

Donald and Violet Logan were in their main deck cabin, putting Helen to bed for the night, when the first torpedo struck. The main lights in the cabin went out, but an emergency light came on, and so at least they could see.

"Violet?" Donald said, as calmly as he could, "I think we've been hit. Best you put on your life jacket and then wrap up Helen as best you can. We must get to our lifeboat right away."

"Oh, yes," Violet said, and quickly grabbed her life jacket, putting it on just as they had practiced at drill, while Donald did the same with his.

The second torpedo hit just as Violet was wrapping the baby up in her blanket. Violet, desperately holding the baby Helen in her arms, then made for the cabin door, only after the doomed ship had once more settled out.

"I'll just grab some nappies," Donald said, did so, and followed his wife and child out of the cabin. They found the passageway outside their cabin filled with other passengers in various stages of panic. The three Logans then made their precarious way as best they could to the port side lifeboat station assigned to them.

* * * * *

Tom Buckingham, walking downhill, and trying not to slip, went to the port side of the ship to see to the lifeboats there. Here the problem was the reverse of what he had seen on the starboard side: when the boat davits were swung out over the side of the ship, the suspended boats were several feet off the rail, and the gap between the rail and the boat became a divide difficult to cross for most of the passengers. Making matters worse, the *Laconia* was now dead in the water, and bobbed up and down with the motion of the surging sea below. The ship's motion, of course, set the suspended lifeboats swinging.

Buckingham watched as several of the passengers, attempting to leave the heaving deck and cross over the gap into the waiting lifeboat, missed their moving, swinging, target, and fell overboard instead.

*****

At their lifeboat station, Donald Logan held onto Violet, as a chivalrous man, already in the boat, reached out and guided Violet and the baby into it. Then Donald boarded the boat, where he and Violet and the baby bunched up, making room for others. Donald guessed there were about ninety people in the boat when the crew finally began lowering it.

Settled in the boat's stern, and one of the first to board the boat, was James Fellows.

The boat was several feet off the ocean surface when the falls hung up, and the boat would lower no farther. One of the crewmen on *Laconia's* deck made to cut the lines with a fire axe, cutting the forward line first.

With the forward falls severed, the lifeboat hung at a crazy angle, its stern in the water, the boat lurching with the heaving sea. Fellows and some dozen other of the passengers were then spilled from the boat, but, with Violet Logan clutching the sleeping baby to her breast, she and Donald managed to hang on. When the bow line was also finally severed, the boat slammed onto the water's heaving gray-blue surface, and several more passengers fell overboard. Once again, the Logans managed to hang on.

James Fellows had struck his head on the lifeboat's transom when he spilled from the boat, and was knocked unconscious. Kept afloat by his life jacket, he found himself well away from both the lifeboat and the *Laconia* when he regained consciousness.

* * * * *

Marco Scarpetti hung back from the surging POWs as they hurled themselves against the cage doors that secured their prison. Willing himself to remain as calm as he could, he took stock of what he knew was a desperate situation.

Upon first arriving aboard the *Laconia*, Marco had surveyed his prison, and learned all he could about it. The hold, he knew, was ventilated by four large intake ducts, two on each side of the ship. These, he had determined, went straight up to the main deck. While large enough to accommodate a person, the ducts were impossible to climb when the ship was on an even keel. Marco knew as well that the air ducts were also closed off from the outside by a grating on their main deck — its purpose being to keep anyone from falling *into* the duct from above.

Now Marco saw that the ventilation shaft was at a sufficient angle that a small and agile fellow such as he was could easily use it to shimmy up to the main deck — and freedom — at least freedom from the *Laconia's* hold. Then, as if the madness taking place in front of him was not enough, he now heard shots being fired. Without further thought, he started up the closest air duct.

Progress up the shaft was slow, but steady, and, as he made his way up the duct, he began to consider the grating at the upper end. *I imagine it's screwed shut,* he thought, *but I might be able to kick it free, if only I could turn myself about in this shaft.* But he knew that such a maneuver was impossible, even for him, because the shaft was simply not wide enough. *I need to get there first,* he finally concluded, *and worry about the grating when I get there. Perhaps it will just lift off.* But this last, he feared, was just wishful thinking.

Marco finally did arrive at the grating, and it did not "just lift off."

* * * * *

Jim McLoughlin had been making his way belowdecks to meet up with the chief steward and get his "gourmet meal," and had just reached the third deck passageway when the nearby blast knocked him off his feet and onto his belly. His fingers scrabbled, trying to grab a hold, as he slid on the steel deck of the wide passageway. He had almost crashed feet first into the port bulkhead when the ship lurched back in the other direction. Only by wrapping his arms over his head did he avoid serious injury, as he was hurtled onto the opposite bulkhead. Regaining his balance, and with his ears still ringing from the blast, Jim made his way to the stairwell he had just descended, only to find that its upper portion — the way out — was now a mass of tangled steel.

But McLoughlin knew the ship. He had roamed every inch of it as a child. He knew the stairwell was backed by a Jacob's ladder that led to the main deck. He was making his way toward it when the second blast occurred. The blast was farther away this time, but the ship again lurched, again to port, and he was again thrown onto the deck.

The passageway was now filling with people, as passengers and crew made their way to the stairwell only to discover, as Jim had, that the way out was blocked.

McLoughlin got up and made his way to the Jacob's ladder, and started climbing, trying to shout above the panicky din in the passageway, "Follow me!" He was well on his way up to the main deck when he noted with righteous satisfaction that several people had heard him, and were climbing up behind him.

* * * * *

Robby Cotton and Ralph Tinsdale, now separated from their other mate, were forward on the ship's port side.

Summoning every ounce of will at his disposal, Robby fought off the terror that gripped the people around him, the same terror that was trying to numb his brain, and forced himself to think. He figured that trying to get into a lifeboat in the midst of the panicked and frightened people fighting to board them, would be a wasted effort. And then he knew, somehow, that Ralph and his immediate survival required them to get into the water and away from the ship before it sunk. He had heard that when a ship went down, it sucked anyone floating nearby down to the bottom with it. And that, he figured, was most likely to happen if they entered the water on this, the port side. But the thought of climbing up the deck to the starboard side, and then scrambling down the side of the ship into the water, seemed like just too tedious and daunting a task. *What if we take another torpedo?* he thought.

"Come on, Ralph," Robby shouted. "Over the side!" Holding his nose with his right hand, and grasping his scrotum with his left, Robby jumped over the rail, and went into the water below, feet first.

As he struck the water, Robby instinctively gasped for air, and, as he plunged underwater, instead of air got a mouthful of seawater, some of which he aspirated. Robby plunged down farther and farther, the straps from the life jacket gripping his crotch, until, at some depth, he leveled off and then shot upward. Or, at least, he hoped he was headed upward and to the surface — disoriented now, he was not at all that sure — and he could only hope that he would not strike the ship on the way up and kill himself that way.

Because of the seawater he had swallowed, he needed to cough — to force the salt from his mouth and the brine from his lungs — but knew he dare not. Just when he was sure he could no longer fight off the need to clear his lungs, he was suddenly back on the surface, first shooting up well out of the water, then settling back. *At least,* he thought, as he spit and coughed and wheezed, *I didn't hit the ship.*

Robby looked around. The life jacket worked just fine; his head was held nicely out of the water. The *Laconia*, now a night-black hull looming above him, was not twenty feet away. *Ralph!* he thought, suddenly realizing that he had not waited to see whether Tinsdale had jumped with him. "Ralph!" he shouted, treading water, propelling himself around in a circle so he could search in all directions. But it was dark, and he saw nobody.

Finally, despite his continued calling, Robby realized Tinsdale was nowhere to be seen or heard. Now, Robby knew, he had to think only about himself. He shed his shoes, and swam away from the ship as fast as his life jacket would allow.

* * * * *

Marco Scarpetti had escaped from the prison hold below, only to find himself once again confined. This time by the steel grating that blocked his access to the ship's main deck. His chances of surviving, remaining alive, and escaping the sinking ship, now seemed slimmer than ever. "Help!" he shouted. "Can anyone help me?"

All of the lifeboats that could be launched, had been launched on that side, and one of the crewmen was making his precarious way aft on the tilted main deck to cut the remaining wooden rafts loose, when he heard Marco's cry for help. "Poor bugger!" he said aloud, when he saw the helpless, pleading face behind the grating of the ventilation shaft. "I'll be back," he said, and then went in search if something to pry the grating loose. He returned moments later with a strip of steel that had served as a section of hatch batten. A few minutes later, he had managed to pry the grating far enough off so that Marco could escape.

"Thank you," Marco said. "I owe you my life."

The crewman just smiled and said, "It's all tickety-boo!" He then left Marco, and continued to make his way aft, his original mission to free the rafts still in mind.

Marco would never see the man again.

Marco then went over the nearby rail, and skidded down the barnacled and sea-grass-fouled side of the ship and into the water. Swimming away from the ship, he worried that he had been scraped by the barnacles and might be bleeding. *Sharks!* he thought.

There were people in the swelling gray water all about him, some with life jackets, floating, others, like him, no life jackets, swimming.

"Over here!" he heard someone shout out of the darkness. "We can take someone over here!" So Marco swam in that direction. Soon, arms reached out for him and pulled him aboard one of the already overcrowded lifeboats.

* * * * *

Once on the main deck, Jim McLoughlin had made his way aft to the gun and his duty station. But when he finally got to the fantail, he found no one there, and saw that, in any case, the gun was inoperable. The ship was sinking, he knew, and now his duty was to save himself. He looked over the side in both directions, port and starboard. The stern was very high, with the tips of both screws well out of the heaving sea. He saw that, off the port side, the water looked *very* far below. *No bloody way I'm going down that way,* he thought, and decided instead to slide down the sloping side of the ship to starboard.

He found a convenient rope that someone before him must have rigged with the same purpose in mind, and lowered himself down, over the side of the ship, and into the water. He swam quickly away from the ship.

There were people in the water all about, desperately seeking to grab onto anyone or anything to help them stay afloat. McLoughlin had to fight off more than one set of desperate hands seeking to grab hold of him. A strong swimmer, he quickly cleared the ship. He could make out a lifeboat, framed against the now darkly pink dusky sky, and made his way to it. It was overcrowded, but he pulled himself aboard anyway, only to find that he was waist deep in water. Someone had forgotten to plug up the drain hole in the boat before it was launched, and it was flooded. *No future here,* Jim thought, and lowered himself back into the water. He swam away with an easy breast stroke, seeking to put more distance between himself and the dying *Laconia*.

* * * * *

Among the last people to abandon the *Laconia* were Tom Buckingham and Stanislaw Kominsky. Buckingham let himself down a rope hung over the port side into the water below. Like Robby Cotton, Stanislaw had jumped into the water. But Robby and Buckingham wore life jackets; Stanislaw had none—*and* he was a poor swimmer.

Buckingham clung to a floating corpse for at least two hours, as a pink dusk slowly receded into darkness. All around him, sharks and barracuda were feasting on the dead, and even some of the living. When a lifeboat passed close by, he swam toward it. He had observed most of the lifeboats being loaded and launched. This one, like all the others, he knew would be overcrowded, but he pulled himself aboard anyway, with some help from the passengers. Once aboard, he noted that this boat, was, for some reason, not as packed as he had anticipated. Discerning the reason for Buckingham's puzzled look, one of the passengers said, "We lost quite a few when the suspension lines broke and the boat went all askew."

"I see," Buckingham acknowledged. "Thank you, Mr. Logan. I'm happy to see that Mrs. Logan and your child are safely aboard as well."

Logan and most of the other passengers in the lifeboat recognized Buckingham as *Laconia's* third officer. Instinctively acknowledging his authority, they quickly fell into line as Buckingham ordered an inventory of the lifeboat's stores and equipment. He soon brought some order into what had been only disorganization, bordering on chaos.

*****

Stanislaw was almost exhausted to the point of giving up, just struggling to stay afloat in the swelling sea. His feet then bumped against something solid. He had managed to stumble upon what was a section of wooden grating floating just beneath the surface. It provided just enough buoyancy to keep him afloat with a minimum of effort. Now, it was past getting dark, and he knew that this night would be a long and sleepless one.

Gene Masters

2

South Atlantic, 12-13 September, 1942

U-156 was still submerged, and Hartenstein was peering through the periscope at the dying *Laconia*, now framed against the pink dusk. "Leo," he said, addressing his second officer standing in the conning tower beside him, "she's in her death throes. I don't think we need waste a third torpedo on her."

Leopold Schumacher just smiled back at his captain. Victory was always sweet.

"And I think her gun is no longer of any concern either," Hartenstein added. "It appears that the ship has been abandoned. I think it's safe enough for us to surface now, and see if we can locate her captain and chief engineer."

"*Jawhol, Kapitän,*" Schumacher responded, and ordered the boat to the surface.

Once the boat had surfaced, Hartenstein asked if there were any radio signals coming from the stricken ship. "Yes, Captain," Schumacher reported, "a distress signal. But the radioman said the signal was very weak, and we have jammed it."

"Very good, Leo," Hartenstein replied. "It would be unfortunate if we were interrupted before we completed our business here."

The U-boat slowly made its way toward the stricken target, which was still barely afloat. Hartenstein noted that several lifeboats had been launched—he could count at least seven—and that there were many people in the water. He headed the submarine toward the closest lifeboat, intent on locating the target's captain and first officer, when someone in the water, and close by, called out to the passing U-boat, *"Aiuto! Aiuto!"*

That's Italian! Hartenstein thought, *an ally!* A cold chill gripped him; was it possible the *U-156* had sunk an ally's ship? "Quickly," he ordered, "bring that man aboard!"

Once the Italian was brought aboard, Hartenstein, who spoke no Italian, had a crewman, who could speak that language, relay his questions. "What flag does the ship fly?" he asked.

"Britannica." The answer required no translation. Hartenstein was greatly relieved to hear that the target was, indeed, British; perhaps the Italian's presence was just a coincidence? But, then, Hartenstein didn't believe in coincidence.

"What is the name of the ship? Why is an Italian aboard?"

"The name I do not know, and I am—*was*—a prisoner of war. One of many."

"How many?"

"Moltissimi. Non sono sicuro, ma migliaia."

The Italian was unsure, but said that there had been maybe "thousands" of Italians aboard.

That, Hartenstein knew, had to be impossible, but there could still have very well been hundreds, rather than "thousands," of prisoners aboard a liner that size.

"Well, Captain," Schumacher asked, "shall we begin to fish as many Italians as we can out of the water?"

"And just let the non-Italians drown? Could you do that, Leo, could you just let the others die, just push them away from the boat, and pull aboard only Italians?"

At first, Schumacher looked confused. Then he ordered all off-duty crewmen topside to begin rescue operations. Once the men were in place, he ordered, "Bring aboard anyone in the water who comes alongside. Tell those survivors in the lifeboats to stand off for now—that we must fish out those in the water first." Hartenstein allowed himself a slight smile upon hearing his first officer's orders.

* * * * *

Hartenstein stood on the U-boat's tiny bridge and watched with satisfaction as his first officer continued to supervise the rescue operation. The crew had begun, as ordered, hauling aboard any person who came alongside the boat, regardless of the language they had used to cry for help, whatever their nationality.

Hartenstein singled out a survivor in a British army uniform and invited him up to the bridge. He soon learned from the Englishman that he had struck the *HMT Laconia*, and that she had been carrying, perhaps, 1,500 Italian POWs.

At 9:11 PM, barely visible against a moonless night sky, the *Laconia's* stern lifted high out of the water, and the doomed liner sailed bow first beneath an inky, heaving sea. Just as her stern disappeared from sight, there was a loud underwater explosion, as *Laconia*, her boilers flooding, responded to the final insult hurled at her from the triumphant deep.

* * * * *

Schumacher saw to it that everyone brought aboard the submarine was made as comfortable as conditions allowed. Men who appeared fit enough, remained on the boat's crowded deck. Women and children, and anyone injured, were brought below. The injured were attended to by the boat's corpsman, and those men whose injuries were deemed as "not serious" were returned topside after treatment.

The boat's cook prepared a thin, hot, gruel and cups of the comforting fluid were passed to all the survivors aboard, including those topside. Eventually, either he, or his relief, would prepare food continuously over the next five days.

When a lifeboat was brought alongside, the passengers were first asked if the ship's captain, or first officer, was aboard. When they were not, women and children were offered the opportunity to come aboard. Some, but by no means all, demurred. Anyone injured was always brought aboard for treatment. All were fed the hot soup. Then the lifeboat was passed a line from the boat's stern, and taken in tow.

When it soon became clear that no more survivors could be accommodated belowdecks, those brought aboard had to remain topside until someone else could be brought up from below; then someone topside could take their place belowdecks. While topside, the survivors huddled together on deck. By 1:00 AM on the

The Laconia *Incident*

13th, *U-156* had rescued all Hartenstein thought his boat could comfortably hold aboard: ninety souls.

Being aboard a lifeboat, or huddling topside aboard *U-156*, was unquestionably better than floating in the sea, but it was still fraught with hardship. With a freshening breeze, sea spray continually wet the survivors down. In the sunny daytime, the wetness would become a salt crust on skin and clothing, but now, in the night, it sucked the warmth from a body and left a person shuddering.

Hartenstein well knew that the solution to the situation at hand was well beyond the capability of his boat and its crew by themselves. The surviving Italians alone, he saw, had to number in the hundreds. At 1:25 on the morning of 13 September, he therefore radioed BdU in Paris, requesting that Vice Admiral Dönitz send instructions.

The dispatch read:

Sunk by Hartenstein, British Laconia, *Qu FF 7721, 310 deg. Unfortunately with 1,500 Italian POWs. So far 90 fished. 157 cubic meters* [fuel oil]. *19* [torpedoes], *trade wind 3, request orders.*

Gene Masters

3

Paris, 13 September, 1942

Vice Admiral Karl Dönitz was awaken from a fitful sleep. A glance at the windup alarm clock next to his bed told him it was 0347. The meticulously uniformed messenger who had awakened him stood at attention as the admiral asked, "What is it, Herbst?"

"An urgent dispatch from *U-156* commander Hartenstein, Admiral!" he replied.

"Give it here," Dönitz said, as he swung is bare legs out from under the covers, sitting up. The air in the room was chilly, and raised goose bumps on his legs. Herbst then handed him the dispatch, and Dönitz quickly read it. It took him only seconds to absorb the contents. "Who has the duty?" he asked.

"*Kapitänleutnant* Zimmermann, Admiral."

"Very good. Tell Zimmerman I want my staff assembled in the operations center in half an hour. See to it!"

"Yes, Admiral," the messenger replied, and hurried from the room.

"Hartenstein," Dönitz said aloud to himself, "now what have you gotten yourself into?"

* * * * *

Karl Dönitz surveyed the room, his steel gray eyes passing from one of his staff to the next. All were, as was he, in full uniform. Some appeared rested, despite the early hour, but most were bleary-eyed, obviously recently awaken from sound sleep.

"Gentlemen," Dönitz said, handing the copy of Hartenstein's dispatch to his chief of staff, and indicating that it be passed around the table. "It seems that *Korvettenkapitän* Hartenstein has put us in somewhat of a pickle. His *U-156* has just successfully dispatched an armed British troop transport, the *Laconia* – which is the good news. It appears, however, that the ship was transporting what appears to be 1,500 Italian prisoners of war—which is the not-so-good news. Hartenstein has begun a rescue operation, doing what he can for both the Italians and the other survivors, but there is a limit to what one U-boat can do. He has asked for instructions. Obviously, we must do what we can, at least for our Italian allies." He paused, and again looked around the table, as if taking each man's measure, then said, "I am asking you for your input. Comments, gentlemen?"

Dönitz's chief of staff, a *Kapitän zur See*, or captain, spoke up. "We have some assets in the area, Admiral, Würdermann's *U-506*, Schadt's *U-507*, and a supply boat, Wiliamowitz-Mollendorf's *U-459*."

"Very good," Dönitz responded. "Dispatch all three to square 7721 at full speed. Any other ideas?"

"What about the Italians? Can't they help?" another aide, a *fregattenkapitän,* or junior captain, asked.

"Or the French, perhaps?" asked another.

"For the French, the Vichy are technically neutral, so we will need Raeder," Dönitz said, referring to his boss and head of the *Kriegsmarine, Großadmiral* (Grand

Admiral) Erich Raeder. "I will call the grand admiral in Berlin first thing—after I call my Italian counterpart. The Italian surface fleet may well be bottled up in the Mediterranean, but perhaps there is a submarine or two in the area that can help us. Get me Admiral Parona on the telephone."

A call was put through to Rear Admiral Angelo Parona, commander of the Italian submarine fleet, at BETASOM, the Italian submarine base, in Bordeaux, France. He quickly agreed to dispatch his closest submarine, the *Commandante Cappellini*, to the area.

* * * * *

In Berlin, later that morning, and before contacting the French for assistance, Grand Admiral Raeder had the unenviable task of informing Hitler of the situation in the South Atlantic. *Der Fuhrer*, needless to say, was not pleased, having already let his feelings be known about enemy ship survivors. On the other hand, he had come to hold Dönitz and his U-boat commanders in high regard. Their successes in the North Atlantic were taking the Third Reich far closer to bringing England to its knees than Hermann Göring's Luftwaffe had done. His portly and pompous air marshall had promised to do just that, to "bring England to her knees," but with no measurable success. He had had to shift his air arm's mission in any case, it was needed in support of the Russian campaign. Now his U-boats were the single most effective weapon against the British and the Americans.

"Do not expose our U-boats to any danger," Hitler ordered, "and see to it that the matter is dealt with, and over quickly." Raeder passed on that exact message on to Dönitz.

Raeder then contacted Admiral Collinet, the French fleet commander in Dakar, Senegal. In explaining the situation, Raeder neglected to note that the survivors included British citizens. The French were still livid over Churchill's decision to destroy much of the French fleet to avoid its falling into German hands. Collinet agreed to dispatch two sloops, the *Dumont-d'Urville,* and the *Annamite,* and the cruiser *Gloire* to the scene.

U- 459's commander, Wiliamowitz-Mollendorf, then notified BdU that he was too far away from the area to reach it in time to do any good. BdU therefore withdrew the order for the supply boat to join the rescue effort.

Dönitz then contacted Hartenstein by radiotelephone, telling him exactly what help was on the way, and when to expect it.

"Tell me, Werner, was an SOS sent by the *Laconia*?" Dönitz asked.

"Yes Admiral. They got a signal off before we could jam it."

"Very well. Carry on, Werner," the admiral replied, and then ended the transmission.

Now, Dönitz worried about his vulnerable submarines and the other allied shipping he had heading to the scene. He shuddered at the distinct possibility that every enemy ship in the area would be converging on square 7721 as well. Nonetheless, he did not withdraw the order for *U-506, U-507,* and the other vessels to join the rescue effort.

Gene Masters

4

Square 7721, Nighttime Sunday, 13 September, 1942

Robby Cotton leaned back on his life jacket, knees bent, arms out, just as he had been taught at *HMS Drake*, conserving his energy. The swelling ocean water wasn't terribly cold, but it was still below his body temperature, and, he knew, was therefore pulling heat from his body. At some point, without food, he would succumb to hypothermia — that is, of course, if he didn't die of thirst first. Or wasn't attacked by a shark — or a barracuda.

Are there barracuda in these waters? he wondered. There were sharks for sure. He had seen them from the deck of *Laconia* only yesterday, swimming lazily alongside. At least, now, any fear he had felt had left him completely — now that death was a distinct, even imminent, possibility, he was somehow no longer afraid.

Robby found himself thinking about God — something he had done very little of since he had lost his family to the Nazi bombs. He hadn't behaved very well since he had joined the Navy. He hadn't really behaved very well even before then. He wondered if God would let him into Heaven, anyway. He hoped He would. And so, Robby prayed.

Robby could see nothing. There was no moon whatever, save a thin crescent, and all was an inky black. There were people around. He could hear them, but could see no one. Despite fighting to stay awake and pray, Robby was just dozing off, when he suddenly found himself bathed in light. It was coming from a grey eminence approaching from the darkness to his right.

"*Hallo, kannst du mich hören?*" a voice called out from the light. "Can you hear me?"

German. The thought flashed through Robby's mind. Answering in English might just well earn him a hail of bullets. *But perhaps that would be better than dying of thirst or being eaten by a shark. Bugger all! But what if it's God what sent them, whoever they are?* Then he thought, *Guess there's nothing for it.* Finally, he cried out, aloud "Over here," answering the disembodied voice. He had half expected a hail of bullets in reply, but minutes later, strong arms were hauling him aboard a German submarine.

* * * * *

With *Laconia* now committed to the deep, *U-156* had been maneuvering in the area, checking on those in the lifeboats, offering hot coffee, or tea, and some food, and pulling still more survivors from the water. This continued throughout the night. Somehow, the Germans had managed to cram another hundred people aboard, and were towing still more in their lifeboats.

* * * * *

Hauled aboard *U-156*, Robby looked around him. He was very close to the boat's bow, and from what he could make out in the dark, there were people everywhere on deck. If they were off the *Laconia*—and they had to be—he should recognize at least some of them. But it was too dark; they were all just gray ghosts. Some were huddled under blankets in groups of three or more, others just squatting on the deck or standing. *What the devil is going on?* he asked himself. *This is the enemy — the same lot of Nazi bastards that murdered my family, and the same U-boat bastards that sank the* Barham. *What kind of devious shite is this? What are these German bastards up to?*

He was given a cup of warm tea to drink by a German sailor, which he downed quickly and, he hated to admit, gratefully. A while later, another German sailor, a young man about his own age with a pleasant smile, came up to him, and said something to him in German, which, of course, Robby couldn't understand. The sailor, however, was gesturing for Robby to follow him, saying, *"Kommen sie,"* which Robby understood well enough. Then, padding after him in his stocking feet, Robby warily followed the man aft, as they weaved their way between the squatters along the crowded deck.

As they passed under and alongside the bridge, where some lights had been strung up, a man, wearing a German naval officer's cap and a cadaverous smile, greeted Robby with "Hello, Royal Navy!"

Robby got the distinct impression that the man was genuinely glad to see him. Despite himself, he couldn't help but nod and smile back at the man. *But he's the enemy!* Robby's mind screamed at him, *one of the bastards that killed my family!*

"*Der Kapitän,*" his escort said, which, again, Robby understood well enough.

His escort took Robby belowdecks, through a hatch well aft of the conning tower, and then led him forward on the submarine. The first thing to hit him was the belowdecks stench: diesel fuel, unwashed bodies, with a peculiar damp, musty, undertone. The first space he entered was crowded with people, and he could barely make his way forward through the boat.

There were people he recognized now—at least some of them—people whose faces were familiar, almost all of whom he had never spoken to: women and children from among the *Laconia's* passengers, and the wounded, mostly men, but some women and children as well.

In the next compartment, he passed by the pulsing diesels. They were draped with drying clothing. He brushed by still more damp clothes hanging from lines hung willy-nilly in the compartment. The heat from the engines, and the steady draft created as they pulled in combustion air from the boat, aided the drying process.

Robby's escort chatted as they made their way forward, speaking pleasant-sounding enough German words, completely unintelligible to Robby. The tightness of the place, the way he could almost reach out and touch the inward-curved bulkheads with his hands, alarmed Robby. And how jammed up the tiny space was with pipes and cables and machinery! It all made him very uncomfortable, uncertain that the tiny space around him would hold enough air for him to breathe.

Still inside the engine room, his German escort stopped Robby, then stepped back, sizing him up. Looking about, the man selected some of the hanging clothing—a shirt, some trousers, warm socks, underwear—and handed it to Robby. The bundle of clothes was warm and dry. It felt wonderful. His escort smiled and again said, *"Kommen sie,"* and then led Robby forward.

Next, they passed through a round, watertight doorway and into what had to be the crew's sleeping quarters, also crammed with women and children, and the wounded, all jammed in close together—the lucky ones in the bunks. Again, some he recognized, others not. Then, Robby was led through yet another doorway, into a crew's mess: narrow tables and benches—same tightness, same oppressive feeling. *How can these men live and work in such a place?* he asked himself.

There, in the mess, another seaman, an older man with a gravelly voice, pointing to Robby's bundle of dry clothes, said in halting English, "You, get naked, put on this."

Understanding he was to get out of his wet clothes, and put on the dry ones he was holding, in front of these strangers, Robby was, at first, somewhat uncomfortable. But then he felt really stupid when he looked around the compartment, and saw other men doing just that: stripping down so as to exchange their wet clothes for dry ones. He then quickly—and gladly—shed his wet clothes.

Whoever had worn Robby's "new" clothes before was obviously bigger than him. *No matter,* he reasoned, *the clothes are dry—and warm. Now to get out of this place and back into the open air!*

Back topside, and settled at his old spot near the bow, Robby, despite the damp darkness, felt grateful to be out of the cramped and stifling spaces belowdecks. He had heard that submarines were prime duty—for others, perhaps. *Them that rides 'em,* he thought, *is bloody well welcome to 'em.*

Toward morning, Robby was given some biscuits and a cup of warm, thin soup, and was grateful for them. However, he was *very* confused. He had expected murder and mayhem from the enemy. But he had *never* expected mercy.

5

Square 7721, Daylight, 13 September, 1942

In the morning, Hartenstein notified BdU that there were 193 survivors aboard *U-156*. He was overjoyed to learn, in turn, that help was on the way, but dismayed that the inbound U-boats and surface vessels might not arrive for several days.

Robby had passed the night alternately standing and squatting on the deck of the U-boat, so close to other survivors that he did not fall over, even when he surprised himself with a newly-realized ability to sleep in either position.

Whenever he was awake, he realized that the Germans had been indiscriminately pulling survivors from the water; on deck around him were conversations in Italian and Polish, as well as in English.

Upon reflection, he realized, *They fished me out of the sea even after I answered 'em in English, so why not anyone else live and afloat? But then I wouldn't 've been in them straights in the first place, would I, if the U-boat hadn't fired two torpedoes into my ship!*

Upon further reflection, however, he decided, "*But that's, after all, their job. It's what U-boats do, sink ships, and our two countries are at war.*

It just didn't made sense. Robby had never, ever, expected humane treatment from the enemy—especially not *this* enemy. He had heard the stories of the atrocities committed by the Nazis on the Continent, and had expected nothing less at sea. Yet here he was, standing high and dry on the deck of a German submarine. And, yesterday, the boat's commander had been downright gracious! It was all very confusing.

The night had been dark black, despite the panoply of glittering stars spatter-painted across the inky backdrop. Now, at dawn, the first glimmering light rose up like a glimmering broom, sweeping those stars away from the eastern sky.

"Robby, mate, that you?" a familiar voice came from Robby's left, not six feet away.

"Jim?" Robby replied. "You're here?"

"Pardon me, mate, pardon me," Jim was saying, as he made his way past several other survivors to Robby's side. "Yes, Minnow, it's me. Have you been standing here all night? And me just a few feet behind?"

"So it would seem, Jim!" Robby could barely contain his joy at seeing his friend, alive and well. "Bugger me, but it's good to see you!" The two men hugged each other in what was definitely an uncharacteristic English fashion.

"Seen any of the others?" Jim asked.

"Not since we were aboard," Robby replied. "Last I saw Fellows, he was pushin' his way into a lifeboat. And then I tried to get Tinsdale to jump with me, but then I realized he hadn't, only not until I was in the water. We'd lost track of Martin earlier."

"The lieutenant?" Jim asked.

"Him neither. Last I saw Tillie he was helpin' the crew load the lifeboats."

McLoughlin nodded, his face drawn. "Sounds like Tillie," he said.

In further comparing notes, Robby learned that McLoughlin had been fished out of the water just as he had, and that his treatment had been identical to his own, right down to the belowdecks tour, the dry clothes, and the biscuits and soup.

"I never thought I'd make it to see this dawn," Jim said. "I was gettin' bloody tired, just treadin' water, and I was sure I was shark bait, if I didn't just drown first."

"Same with me," Robby agreed. "And I was sure, when they found me, the Germans would shoot me where I swam."

"And me. Why, the U-boat captain was downright cordial! Greeted me after I was hauled aboard, asking me if I was hungry, and even offered me the food off his plate! I'm thinkin' to myself, 'Jim, lad, isn't this the same bloke what just lobbed two torpedoes into your ship?' And I had no answer to that, 'cause here he was now, actin' the perfect gentleman."

"And so he seems," Robby agreed. "But, somehow, I keep waitin' for all to go sideways!"

"I know what you mean, Minnow. I know what you mean. Somehow, it's all off kilter from the way you'd expect. It's like you said, and I'm waitin' for that other shoe to drop!"

But little did the two friends suspect that it was not the U-boat commander who would drop the other shoe.

* * * * *

Hartenstein ordered his crew to start taking the lifeboats in tow just after sunrise that morning. As they were brought alongside, Hartenstein noted that in one boat in particular, the passengers were in some order, with women and children seated, and men standing or on their haunches in the well deck. Not a single passenger, unlike in the other lifeboats, was in apparent panic. He knew at once the reason for it.

"May I enquire as to the officer aboard?" Hartenstein asked, calling out in English to the boat from his perch on the U-boat's bridge, not fifteen feet away. Tom Buckingham stood up. "And you are, Sir?" Hartenstein inquired.

"Thomas Buckingham, Sir, lieutenant, Royal Merchant Navy, and third officer late of *His Majesty's Transport Ship Laconia*."

"I am pleased to meet you, Mr. Buckingham," Hartenstein replied. "I am Captain Hartenstein, and I do apologize for sinking your ship. But it is a war, after all, is it not?"

"It is indeed, Sir," Buckingham called back.

"I regret that we are out of dry clothing for the moment, but my men will take you in tow and give you and your people some tea and something to eat."

"Thank you, Captain," Buckingham replied, and meaning it.

"And I see there's a young woman with a baby aboard," Hartenstein observed. "She is welcome to come aboard straightaway and go belowdecks. I'm sure she and the child will be more comfortable there."

Violet Logan, upon hearing Hartenstein's offer, spoke up. "Thank you, Captain, but I would prefer to stay with my husband."

"As you wish, Madam," Hartenstein replied, touching the rim of his cap.

The Laconia *Incident*

* * * * *

The lifeboat carrying Marco Scarpetti had drifted off during the night and become separated from the others. Also during the night, two passengers, one Englishman, whose arm had been ripped open, and a fellow Italian, who had been shot, died. At first light, some of the other passengers said a few prayers, and their bodies were rolled over the side with as much dignity as such a maneuver would allow.

Later in the morning, the lifeboat passed close aboard a man who seemed to float straight up, only his shoulders and head showing. His eyes were shut, and everyone in the lifeboat assumed he was dead, *until* his eyes opened wide, and the man stood up straight in the water, the water then covering only his legs. He shouted something unintelligible, and waved his arms.

"Stanislaw?" Marco called out, recognizing his friend.

"Marco!" Stanislaw called back.

Using oars, two of the men in the boat awkwardly maneuvered it close to the man in the water. It was only when they heard the "clunk" as the boat struck the underwater grating that had supported Stanislaw throughout the night, did they understand his strange ability to stand upright in the sea. Stanislaw was able to then simply walk two steps to the boat, and enter it. Marco's clothes had dried somewhat during the night, but were quickly wet again as he hugged his sea-soaked friend.

"Stanislaw!" Marco kept repeating as he embraced his friend. "How good it is to see you!"

*****

It was still early morning that Sunday when a frustrated Hartenstein did something unprecedented. He was unsure as to when any help would arrive, and was dismayed by the number of people still in the water. He was also somewhat concerned that the Italians in the lifeboats might decide to seek revenge over their former captors. And so, Hartenstein decided to send a message out over the international emergency frequency in English and in the clear:

If any ship will assist the wrecked Laconia *crew, I will not attack her, provided I am not being attacked by ship or air force. I picked up 193 men. 4°- 50" South, 11°- 26" West – German submarine.*

The British in Freetown, Sierra Leone, received the message, but didn't credit it, assuming it was some Nazi trick. The French at Dakar also heard the signal, and also figured it was some sort of German ruse. The Americans on Ascension Island heard nothing.

Aboard *U-156*, Hartenstein ordered his men to paint a red cross on a white bed sheet, and to then drape it over the forward gun mount. The banner was in place shortly after noon that morning, Sunday, 13 September.

6

Square 7721, Monday and Tuesday, 14-15 September, 1942

It was late Monday morning when *U-506* arrived on the scene. Her captain, *Kapitanleutnant* Erich Würdermann, could see that *U-156* had at least 100 men on her narrow deck, and a half-dozen lifeboats in tow, each one filled to the gunwales with survivors. He drew his boat close aboard to *U-156*. Hartenstein, who had slept very little the night before, was on the bridge, and greeted his counterpart, calling out in German, "Erich, my friend, how goes it?"

"Never mind that, Werner," Würdermann shouted back in the same language, his thin lips in a broad smile, "What kind of a mess have you gotten us into now?" Würdermann was hardly an example of Hitler's pure Aryan. He was short and dark, with brown eyes and slicked-down black hair that he parted in the middle.

"It is indeed a mess," Hartenstein acknowledged, "but I couldn't just sail away and do nothing—especially since the greater part of the survivors are our allies!'

"So I understand."

"I am concerned that our Italian allies aboard will take it in their heads to revenge themselves against their former keepers. Would you be willing to take all the Italians on board my boat onto yours?"

"If you wish. Have you have seen no other ships nor aircraft?"

"*Nein*, none whatever.

"Do you think, if any enemy ships or aircraft come, that your red cross banner will stop them from attacking?"

"One can only hope," Hartenstein replied, "that they would respect international law."

Würdermann shrugged. "Well, let's get to it, then."

With that, the crews of the two boats began the business of transferring all the Italians aboard *U-156* to *U-506*. First, the two boats tied up alongside each other. The sea was relatively calm, with gentle swells, the two boats moving together in almost perfect rhythm. Even so, most of the former POWs were frightened, and reluctant to leap from one deck to the other. None-too-gracious prodding by the seasoned submarine sailors was, therefore, required to move the operation forward. Even so, the transfer took well over an hour to complete. When the transfer was complete, Würdermann had ninety-three Italians aboard.

When the boats were finally uncoupled, Würdermann maneuvered his boat well apart from *U-156*. Once Hartenstein's boat was out of sight, *U-506* began to roam the area, setting about the task of fishing other survivors from the water. Würdermann also took aboard women, children, and the injured from whatever adrift lifeboats *U-506* came across. Soon it made little difference that the Italians had been transferred aboard his boat, since Würdermann could no more rescue just Italians, than Hartenstein could earlier.

By day's end, *U-506* had over 200 survivors aboard, and *U-156* was well out of the range of tactical radio. *How long before the French arrive?* Würdermann wondered. *How long before I can be free of this mess?*

The Laconia *Incident*

As the situation continued to develop, it turned out that Hartenstein had been worried over nothing. Not one of the former POWs ever took it in his head to seek reprisals against his former captors. The Italian survivors, it seemed, were just grateful to have been spared, and to be in safe hands.

* * * * *

The lifeboat containing Marco and Stanislaw had become separated from those nearby and had drifted off. Most of those aboard were English passengers, including a smattering of British army and RAF troops. And many of those had been wounded in battle, and were aboard *Laconia* for passage home. Marco now counted eighty-two souls in the boat, including Stanislaw and himself. Another passenger, he knew, had died the previous night, and his body had been let gently over the side.

At least two or three others were in a bad way, and probably wouldn't survive another day without skilled medical attention. By the third day after the sinking, two of the three had died, and were also committed to the deep. Everyone still alive aboard the lifeboat was dehydrated and listless. The previous two nights had been cold and wet (especially wet for those, who, like Stanislaw, had been fished from the water), and the days under the tropical sun were hot and oppressive. There was a little fresh water aboard, and some tins of biscuits; the senior military man, an RAF 1st Lt., had taken it upon himself to ration and distribute those, and no one had argued.

There was also a compass, and most aboard knew that Africa was somewhere off to the east. But, since nobody knew exactly where they were, or how far off landfall was, the consensus was that setting the sail, or rowing the boat, was pretty much useless. The unspoken resolve of the majority was to tough it out and wait to be rescued. Marco wasn't sure he agreed, but was just too tired to come up with an alternative plan.

* * * * *

In the early afternoon of Tuesday, 15 September, *U-507*, *Korvettenkapitän* Harro Schacht commanding, arrived on the scene. Schacht was dismayed with the situation in which he found *U-156,* as was Würdermann.

Schacht, in a rubber boat launched from the deck of his U-boat, came alongside *U-156*. The sun was high in the sky, there was no breeze whatever, and the mirror sea was dead calm. The survivors crowded on the deck of *U-156* somehow parted, making way for Schacht, as he was pulled aboard to join Hartenstein on the U-boat's bridge. The men who had rowed their captain over, pushed off, letting the rubber boat drift away and stand off from *U-156*.

"I can only wonder how Dönitz convinced *Der Fuehrer* to go along with this," Schacht said, as he greeted Hartenstein. Unlike Würdermann, Harro Schacht was light-skinned, blue-eyed, and sandy-haired, the very model of Hitler's super race.

"Well, Schacht," Hartenstein replied, "I don't imagine the admiral asked *Der Fuehrer* for permission.

The Laconia Incident

He probably said that we had no choice, since hundreds of our Italian allies were among the survivors"

"That is true enough. But we are also rescuing Englishmen and Poles. Hitler's feelings on that point are well known. I'm sure he would have us pulling only Italians out of the water and shooting the rest."

"He might at that, Schacht," Hartenstein agreed with a tight smile, "but thank God the *Kriegsmarine* is not the SS. I could never execute such an order."

"Be careful, Hartenstein. I would never say anything, but you could be shot for less."

"Then so be it. And you, Harro, would you shoot enemy survivors?"

"Not a choice I have to make, fortunately," he replied, dodging the question. "The admiral's orders are to assist you in the rescue, and turn over all survivors to the French when their ships arrive. And I will obey orders. My only worry is that an enemy warship or aircraft will not understand the nobility of our mission here."

Schacht was not reassured when Hartenstein told him about his Sunday broadcast over the emergency frequency.

* * * * *

That same Tuesday, the British Admiralty at Freetown sent a message to the Americans at Wideawake Field on Ascension Island.

The message was somewhat garbled, but it *did* alert the Americans that a passenger liner had been sunk with some 700 passengers aboard. It gave the approximate coordinates of the sinking, but they were incorrect, and much closer to Ascension Island than the actual sinking site. The dispatch did say that a British rescue ship, the *RMS Empire Haven*, was en route to the scene. It also requested that aircraft be sent to pinpoint the location of any survivors, and to relay that position to the *Empire Haven*. If possible, the planes were to stay on the scene and provide air cover for the rescue.

No mention was made of Hartenstein's transmission, or of any Axis submarines involved in a rescue over the past three days. The message did state that Vichy French warships had left port and were headed to the area.

In response, the squadron commander, Capt. Richardson, sent out a flight of B-25 Mitchell bombers to search for the survivors. But the B-25's 1,350-mile range meant that the flight had to turn back and return to base before coming anywhere near the actual disaster site.

Again, in response to the message, and not understanding that the oncoming French warships were actually a part of a far larger rescue effort, the base commander, Colonel Ronin, put the Army troops at Wideawake on full alert. He feared an imminent French naval barrage, and a possible invasion by enemy troops.

7

Wednesday, 16 September, 1942

The Italian submarine Commandante Cappellini, Capitano di Corvetta Marco Revedin commanding, arrived in the area on Wednesday, 16 September. The sun was still below the horizon, but was ever so gently lighting up the scene in subdued reds and glowing yellows. The wreckage and the survivors were, by that time, spread out over several square miles. It was in that lifting darkness that the *Cappellini* first came upon a lifeboat. The evening before, that boat had fished an exhausted James Fellows from the water.

"Any Italians aboard? *Ci sono Italiani abordo?*" Revedin inquired of the passengers in that first boat, in both English and Italian.

"No. No Italians," came back the reply from the boat, which had fifty-two people aboard, all men. Revedin inquired as to their situation, and one of the men replied they had a compass, a map, and a hand-powered radio transmitter aboard, but that they could use some water. Revedin passed them over some tins of water, and two bottles of wine.

At first light, *Cappellini* came upon a second lifeboat. In addition to men, this one held women and children. Once again, there were no Italians aboard. Revedin offered to take the women and children aboard, but none of the women would leave the lifeboat, and so both they and the children stayed. Revedin then asked if they needed anything, and they replied, "blankets, food and water, a map, and a compass." Revedin had no blankets to spare, nor map nor compass to give them. He left them with a supply of warm broth, some biscuits and chocolate, and some cigarettes.

Revedin then intended to take *Cappellini* in search of *U-156*, but instead came upon more lifeboats, this time mostly full of Italians. In bringing his countrymen aboard, he could no more abandon the others in the lifeboats than could his German counterparts. He soon had his boat filled with survivors of all nationalities, and was struggling to keep a number of lifeboats together. Now Revedin too, anxiously awaited the arrival of the French.

*****

About the same time that the *Cappellini* came upon the first lifeboat, the Americans at Wideawake Field received a second message from the British at Freetown stating that a second vessel, the armed British merchant ship *HMS Corinthian*, was also on its way to the site of the sinking to search for survivors. Again, the Americans were asked to pinpoint the location of any survivors, and to relay that position to the rescue ships. Once more, if possible, they were to stay on the scene and provide air cover for the rescue.

Since the earlier flight had found nothing, Capt. Richardson correctly reasoned that the disaster scene had been closer to the African coast than had been first reported. He therefore decided to commandeer the only aircraft capable of reaching the actual location of the sinking, and returning to base: 1st Lt. James Harden's B-24 Liberator bomber. The field maintenance crew had just completed repairing the problem that caused Harden to land at Wideawake in the first place, and the ship was once again airworthy.

Harden and his crew, navigator 1st Lt. Jerome Perlman, and bombardier 2nd Lt. Edgar Kellar, were hauled out of bed at first light on the 16th, and called into Capt. Robert Richardson's office. Trudging to Richardson's office below the radio tower, the three men were still far too sleepy to admire the same brilliantly awakening sky that was to greet the *Cappellini*.

Richardson outlined their mission. "Jim, I'm going to have to commandeer your aircraft. You, Jerry, and Ed, are now temporarily under the command of the First Composite Squadron—that is, you are temporarily under my command."

"Yes, Sir," Harden replied, never questioning Richardson's authority. "What do you need us to do, Captain?"

Richardson then explained the mission.

At seven o'clock that Wednesday morning, Harding launched his plane from Wideawake and headed in the general direction of the coordinates originally given by the British. The weather was clear, and visibility unlimited. When Perlman confirmed that the aircraft had passed over the original coordinates, Harden continued flying the plane on the same course toward the African coast, just as Richardson had directed.

Gene Masters

8

Wednesday, 16 September, 1942, 4º 50" South, 11º 26" West

The weather had been monotonous in its sameness; the sky overhead was a brilliant blue, the occasional wispy cloud almost motionless. The sea below was a sheet of blue plate glass. At nine thirty, after a two-and-a-half-hour flight out of Wideawake Field, Harden's Liberator bomber sighted *U-156*.

Seconds earlier, a lookout aboard the U-boat had spotted the plane, and alerted Schumacher, who was on the bridge. Schumacher looked in the direction the lookout was pointing, and, raising his binoculars, quickly located the approaching aircraft. "Captain up!" Schumacher shouted down the open conning tower hatch.

Less than a minute later, Hartenstein was on the bridge. As the aircraft approached, and began a long, wide, circle overhead, Hartenstein saw through his binoculars that its markings identified it as American. Schumacher, rifling through a hastily-retrieved identification manual said, "It's an American Liberator bomber — B-25 — long range. Must have come in from North Africa, probably searching for survivors. My guess is that the transport's distress signal must have gotten out after all."

"That could well be the case," Hartenstein agreed, "and now we shall see how the Americans react to what they see."

On deck and in the lifeboats, some of the survivors, seeing the aircraft, began shouting and waving their arms. Among those was Robby, but not McLoughlin. "What do you think, Jim?" Robby asked. "Good, eh? Maybe our people will come to rescue us?"

"We'll see, Minnow, we'll see," McLoughlin said, warily.

Hartenstein said nothing more at first, just staring at the plane through his binoculars. Then he called out to the men standing on the deck, "Stand back from the gun, there. Make sure the Americans can see the red cross!"

In the aircraft, Harden identified the submarine he saw on the surface as German, an identification confirmed by both Perlman and Kellar. Still, the scene below was bizarre. As he circled, Harden and his crew saw the surfaced U-boat, its deck crowded with people, with six crowded lifeboats in tow. The U-boat was heading toward two other lifeboats. As he continued to circle, the people on the submarine's deck moved away from its deck gun, and, draped across the unmanned gun, was a red cross banner.

Ever wary, Harden radioed the sub, requesting that it identify its nationality. He received no reply. Harden reasoned that the submarine either had not heard the message, or had chosen not to reply. He also may very well have, he knew, sent the message on a frequency the sub's radio room was not monitoring.

He then dived his aircraft on the boat for a closer look. The engine noise deafened, and scared the wits out of the U-boat's crew and the survivors alike. On the deck of the U-boat, Robby stood frozen; McLoughlin made ready to dive over the side should the plane begin strafing. He saw that the bomb bay doors were shut, and knew that there would be no bomb released, at least not on this run.

Equally cautious, Hartenstein said to Schumacher, "Leo, order that the anti-aircraft gun crews be at the ready. Tell them to stay close to their mounts, but to stay down and out of sight." As Schumacher relayed his order, Hartenstein added, "And send a crewman aft with a fire ax. Tell him to be ready to quickly cut the towline to the lifeboats." And again, Schumacher made it so.

Hartenstein was still gazing up at the retreating aircraft from the bridge, when an RAF Lieutenant Colonel approached the bridge. "Captain," the man said, "if you give me access to your radio, I can send an English message in International Morse Code to the Americans, tell them that there are British survivors aboard, and not to attack us."

Hartenstein considered the offer for a second, and then said, "Very well, Colonel, I will have a man escort you below to the radio room."

The lieutenant colonel was taken belowdecks, and sent the promised message in International Morse Code to the airplane circling above. Aboard the aircraft, Harden and his crew received and heard the message, but none of the three could read Morse Code.

Harden was now running out of options. A half-hour had passed, and he could only circle a little bit longer before he had to turn his plane around and head back to Wideawake, lest he run out of fuel. He finally decided to radio Squadron Headquarters at Wideawake for instructions. But he was out of radio range with HQ, and had to turn this plane around and head back in the direction of Wideawake.

On the deck of the U-boat below, Robby, Jim, and the other non-Italian survivors were dismayed, afraid that the Americans might be deserting them.

"Composite Base, this is Two-One-Two Zebra, over," Hardin sent over and over, as he flew the bomber southwest.

Finally, there was a response. "You have Composite, Zebra, over."

"Zebra needs direction from Composite Command, over."

"Wait one, Zebra." The radio operator then sent word for Capt. Richardson to please come up to the radio tower. His office was just below, and Richardson was in the tower in just over a minute. He manned the microphone on his arrival, speaking directly to Harden. "This is Composite Command. Go ahead, Zebra."

Harden then described the scene at the rescue site to Richardson, sparing no detail.

"And you're sure it's a Nazi submarine, Jim?"

"Yessir, Skipper, I'm sure."

"And there are people on the deck?"

"Yessir, lots."

Could be that there was some mechanical problem with the sub, Richardson thought. *And the Germans are ready to abandon ship if they can't fix it or help doesn't come soon. And that red cross must be just a ruse — they're buying time, just hoping we won't attack.* Still unsure, however,

Richardson called in his superior officer, Colonel Ronin, and explained the situation to him. But Ronin's thinking only reflected Richardson's.

"If it's a German U-boat, then what's the problem? Attack and sink it," Ronin said.

"But all those people on deck, Sir, and the red cross flag," Richardson had countered.

Now angry, Ronin argued, "It's got to be a Nazi trick, Captain. Sink the bastard!"

His original analysis now confirmed by his superior officer, Richardson got back on the radio to Harden. "Two-One-Two Zebra, this is Composite Command, over."

"This is Zebra, over"

"This is Composite. Sink sub at once. I repeat, sink the sub, over."

"Yes, Sir. Understand sink sub at once. Zebra, over."

"Affirmative, sink the bastard!" Richardson reiterated. "Out."

With the radio link to Wideawake severed, Harden then spoke into the aircraft's intercom, above the engine noise, to Perlman and Kellar, as he reversed the aircraft's course to again close the rescue site. "Tally-ho boys! We're going to get us a Kraut submarine!"

Gene Masters

9

Wednesday, 16 September, 1942, Air Attack

The American bomber had returned to the skies overhead, and Robby and Jim were, at first, elated.

Then, Jim McLoughlin saw the bomber's bomb bay doors open, just as the plane, which had been circling overhead, went into a tight turn to line up with the slow-moving submarine. Robby hadn't noticed the bay doors opening at all, nor had many of the others standing on deck or in the towed lifeboats. Both Hartenstein and Schumacher definitely had.

"Quickly, Robby, over the side!" McLoughlin shouted.

"Wh-wh-a-a-a-t?" Robby stammered as McLoughlin pushed him off the deck and into the water.

"Swim!" he shouted, as he himself dove into the water. Nobody else on deck, nor in the lifeboats, had thought to follow suit—not yet, anyway.

When Hartenstein saw the bomb bay doors opening, he quickly signaled to the crewman aft to cut the towline, and shouted to Schumacher, "Dive the boat, Leo!"

In response, Schumacher screamed, "ALARM!" down the hatch, ordering the boat to dive. The man aft severed the towline with a single stroke, then dove down the after torpedo room hatch, shutting it behind him. But *U-156* had just initiated its dive and was still on the surface when the bomber began its run.

The plane came careening in at full power, its four engines screaming. Stunned, few of the survivors on deck thought to follow Robby and McLoughlin over the side, nor did many of those in the towed lifeboats abandon them. Second lieutenant Kellar, positioned at the bombsight in the bubble under the bomber's cockpit, released two bombs. Both fell wide of their mark, but sent geysers of seawater up into the air. When the bombs exploded, anyone on deck who had doubted the aircraft's intentions were now convinced — the Americans meant to sink their benefactor. Only then was there any move among the survivors on the submarine's deck, or in the now-adrift lifeboats, to "abandon ship."

The submarine was mostly on the surface, with only its bow starting down. Hartenstein had ordered, and the boat was beginning, an evasive turn. Then the bomber began its second run. By that time, everyone who had been on deck was in the water, and those able to do so were swimming frantically away from the boat. Robby and McLoughlin were well ahead of them, thanks to the latter man's quick thinking.

Those still in the lifeboats were at their wits end; abandoning them was out of the question for many, such as Violet Logan and her child, and they could do nothing but sit still in utter panic as the drama played itself out. And, of course, Donald Logan could never abandon his wife and child, and franticly held onto both of them as he glared, powerless yet defiant, at the approaching aircraft. Tom Buckingham had also elected to remain aboard the lifeboat, having felt responsible for the craft and its remaining passengers. He grabbed a paddle and was rowing, in a futile attempt to clear the bomber's flight path.

On its second run, the bomber released two depth charges over the submarine. Kellar's timing was again off—one was released too early, the second too late. Both missed *U-156*. The lifeboat in which the Logans and Buckingham were riding managed to escape unscathed. But the second depth charge hit the lifeboat third farthest from the submarine; it was a lifeboat full of Italians. The depth charge smashed through the boat, exploding below it, and killing all aboard. Yet another lifeboat immediately nearby was capsized, and its bottom ripped out. In terms of lives lost in the action, this attack proved the costliest; British, Poles, and Italians all suffered losses.

On its third run, the Liberator released two more depth charges over its target. One missed entirely, but the other exploded aft and under the stern of the diving submarine, lifting the stern from the water momentarily. Now in the conning tower, Hartenstein began receiving damage reports. When it was reported that the boat was taking on water, he aborted the dive. He then unbuttoned the hatch above him and returned to the bridge. He was considering how to fight back, using his anti-aircraft guns, when he gratefully watched the bomber turn to the southwest and clear the area, apparently bereft of its bomb load.

Well away from the action, Jim McLoughlin and Robby Cotton treaded water and watched the events as they unfolded before them. "That was a close-run thing," Jim volunteered.

"It was that," Robby agreed. "I thought the bloody Americans were supposed to be on *our* side! What now?"

"Now we stick together and wait."

* * * * *

It was the third bomb run that took its toll of the *U-156*. Sea water breached the battery compartment, and several battery cells were releasing chlorine gas. While the crew was able to don gas masks and repair the damage, there were only enough gas masks for the crew, so none of those rescued, and who were belowdecks, could remain there. All were brought up on deck. The deck again became so crowded that Hartenstein was forced to order able-bodied men into the water—first the British and the Poles, and then, finally, the Italians.

Meanwhile, those in the surviving lifeboats fished aboard all those they could. Somehow, two lifeboats approached the spot where Robby and Jim were treading water at the same time. Each one picked up one man. And so it was that Robby Cotton and Jim McLoughlin were brought aboard different lifeboats, and became separated once again.

Soon, all the lifeboats in the area had taken on all the passengers they could hold, but there were still dozens of survivors in the water.

* * * * *

"We have to clear the area, Captain," Schumacher said to Hartenstein. "The Americans could return at any time."

"They very well could," Hartenstein agreed. "But you know we cannot dive the boat until all the damage is repaired." (Hartenstein knew where this conversation was going, and he did not like it.)

"All the more reason to be somewhere else, should they return." Schumacher countered, confirming Hartenstein's assumption.

"There are still people on deck—almost a hundred, I would guess," Hartenstein mused aloud to his first officer. "Those likely to survive in the water are already overboard. Those still on deck could never survive in the water, and we cannot take them below because of the gas."

"Yes," Schumacher agreed, following his captain's line of reasoning. "They could not remain on deck if the boat was to leave the area at any semblance of speed. And running the engines at speed will help clear the boat of the gas."

"And we cannot run the diesels disconnected from propulsion, just to charge the batteries, while the batteries are still compromised," Hartenstein said. "I am well aware of all that, Leo."

"Yes, Captain," Schumacher said, cowed.

Finally, Hartenstein said, "Very well, Leo, there's nothing else we can do. Clear the deck. Order the rest of those poor souls into the water. Then take us up to full speed and we will head northeast on the surface, and clear the area."

"Yes, Sir," Schumacher replied, and reluctantly executed his commander's orders.

Hartenstein went below, not having the stomach to watch being done what must be done. Thus it was that the final ninety-two people aboard *U-156* were ordered overboard and into the water. Very few survived.

* * * * *

Once clear of the area, the crew of *U-156* worked diligently to repair their stricken craft.

After several hours, the repairs completed, Hartenstein radioed Dönitz and apprised him of the latest events. He reported that the repairs were well enough along so that the boat could return to its original mission. Dönitz ordered Hartenstein to " . . . *take no further part in the salvage mission.*"

* * * * *

Harden had turned his aircraft to the southwest and cleared the area, more concerned about having enough fuel to make it back to Wideawake Field, than being bereft of his bomb load. Busy flying the plane, he had not observed the action directly. He had to rely on what Kellar and Perlman saw, or *thought* they saw, as the attack on the U-boat unfolded.

And so it was that, on return to Ascension Island, Harden reported that the targeted submarine had been observed turning over in the water, while its surviving crew clung to wreckage, or were seen swimming toward the lifeboats. Harden and his crew were convinced that they had successfully sunk the U-boat.

Richardson was delighted. He ordered Harden and his crew to turn in for the night, and to return to the scene of their successful action the next morning and resume the search for survivors.

Gene Masters

10

Thursday, 17 September, 1942

At first light, the B-24 Liberator bomber hurtled down the runway at Wideawake Field, and, once airborne, set a course to the northeast.

* * * * *

While *U-156* had cleared the area, and had resumed its patrol, the remaining three submarines, operating over the still-widening area in which the survivors had become dispersed, continued their rescue efforts.

Dönitz' aides had wanted the admiral to cease the rescue operation altogether, but he insisted that the operation, once begun, had to be completed. The French ships were due to arrive later that Thursday, and there were the Italian survivors (their allies after all) to consider. He sent a message to both Schacht and Würdermann that they were under orders to be on the alert for any enemy air activity. They were to transfer all non-Italians to lifeboats, and were to be prepared to dive at the first sign of danger.

Dönitz also relayed the alert for enemy air activity to the *Cappellini*, via the Italian Royal Navy Command.

* * * * *

The Liberator returned to the scene of its previous day's "victory" late that Thursday morning. Harden began to fly a grid pattern, and he, Perlman, and Kellar searched for survivors. There were, they noted, several lifeboats still afloat, and some swimmers still in the water. Harden contacted Wideawake and radioed back that they had found survivors and gave their position. Wideawake, in turn, vectored the two oncoming British rescue ships to the area. The Liberator had sufficient fuel to remain in the area, so Harden continued the search.

Just before noon, Perlman was the first to sight *U-506* cruising on the surface. He got on the intercom to Harden. "Jim," he said, "check out our four o'clock. It's another damned U-boat!"

"Holy crap!" was Harden's response.

The crew of the Liberator couldn't believe their luck. There was no hesitation this time. Harden immediately initiated the attack. "I'm opening the bomb bay doors, Kellar. Get ready to bag us another Nazi submarine!"

But Würdermann's lookouts had spotted the aircraft well before Perlman saw the U-boat, and *U-506* had already cleared the deck and commenced its dive. The boat was well out of danger when the Liberator released depth charges over the place where it had been last seen. The U-boat escaped unscathed, with some 100 survivors aboard.

Harden reported a "possible U-boat sinking" to Wideawake before turning the bomber back to the southwest and clearing the area.

(The record shows that Harden and his crew were later decorated for having destroyed at least one enemy submarine, and probably damaging another.)

The Laconia *Incident*

* * * * *

Earlier that same Thursday morning, *U-156*, now back on patrol in the South Atlantic, received word from Paris that *Korvettenkapitän* Werner Hartenstein had been awarded the Knights Cross of the Iron Cross.

In the afternoon (and again a few days later), Dönitz sent a message to all U-boat commanders. It was later to become known as the Laconia Order. It read:

> 1. Every attempt to save survivors of sunken ships, also the fishing up of men and putting them in lifeboats, the setting upright of overturned lifeboats, the handing over of food and water have to be discontinued. These rescues contradict the primary demands of warfare, especially the destruction of enemy ships and their crews.
> 2. The orders concerning the bringing in of skippers and chief engineers stay in effect.
> 3. Survivors are only to be rescued if their statement is important to the boat.
> 4. Stay hard. Do not forget that the enemy did not take any regard for women and children when bombarding German towns.

"Well, what do you think about it, Leo," Hartenstein asked his first officer after Schumacher brought him the dispatch of Dönitz' order in the U-boat's tiny wardroom.

"It's well deserved, Captain," he replied.

"Thank you, Leo, but I wasn't referring to the Knights Cross, but to this order—you have read it, of course."

"Yes, Captain, I have." Schumacher hesitated.

"Well?"

Finally, Schumacher said, "I think that the admiral is responding to pressure from up above, Sir. I suspect that the general staff and *Der Fuehrer* were very displeased with the events of this week past."

"Yes," Hartenstein agreed, "I think you are correct. But if I am reading between the lines correctly, I don't think Dönitz agrees. I think he still approves of our efforts to behave civilly, despite our almost being blown out of the water by the Americans."

"And why do you say that, Captain?"

"Read the dispatch carefully, Leo, especially the last statement."

Schumacher reread the order. He reread: *"Do not forget that the enemy did not take any regard for women and children when bombarding German towns."*

"But it's true, Captain, the enemy *have* been bombing German civilians," he opined.

"Yes, they have. But the Admiral also knows that we Germans were the first to initiate such action. It was our *Luftwaffe* that bombed English civilian populations beginning in 1940, and continued to do so, right up until our *Panzers* rolled into Russia, and our aircraft were needed for ground support. And weren't they our Zeppelins that first bombarded London in the last war?"

Schumacher looked pensive. "You're saying..."

"I'm *saying* that Dönitz is subtly reminding us that it is we who first breached the standards of civilized warfare, and that it is no wonder that now the enemy responds in kind. I think he is saying that our past barbarities are no justification for further barbarities." He paused, then continued. "No, Leo, I think that the admiral issued this order only because Raeder and *Der*

Fuehrer ordered him to do so, and this last bit is his way of showing his disagreement without getting stood up against the wall!"

Schumacher raised his eyebrows and tilted his head, but said nothing. *Perhaps the captain's analysis is on target — but then again, perhaps not.*

* * * * *

Later that same Thursday afternoon, the French ships arrived in the area where the *Laconia* had gone down, and there rendezvoused with the *U-506, U-507,* and the *Cappellini.* First, the cruiser *Gloire* relieved the submarines of their passengers (the *Cappellini*, however, retained all the British officers it had aboard).

Afterward, the *Gloire,* along with the two sloops, the *Dumont-d'Urville,* and the *Annamite,* began searching the area for the lifeboats the submarine commanders had said were still out there, scattered about the area.

The lifeboat that held Marco Scarpetti and Stanislaw Kominsky was the second one that the *Gloire* encountered. As with their first encounter, this boat contained British, Italian, and Polish survivors.

Once aboard their ships, the French first served the survivors a thin gruel, and attended to those who needed medical attention. The British and the Poles were then separated from the Italians.

Aboard the *Gloire,* Marco at once protested, pointing out to the Vichy *lieutenant de vaisseau* (lieutenant) in charge, in excellent, if accented, French. "My name is Marco Scarpetti, I am a sergeant in the Royal Italian Army. This Pole, Stanislaw, is my friend. He was kind to me and the other Italians, and has never hurt anyone. And the lieutenant," he pointed out the RAF first lieutenant who had taken charge, organizing and supervising the survivors, and rationing their meager supply of food and water, "the lieutenant organized us in the lifeboat, and kept us alive."

"You speak French!" the Vichy lieutenant replied.

"Yes, Sir," Marco answered, "and English as well. But I was telling you—"

"I know, I heard you. The Pole is your friend, and the Brit saved all your lives. But what do you think we plan to do to them? We French are not barbarians! Still, they are on the other side—they will be treated well and fairly, but they will be kept confined while here aboard the *Gloire,* and in Dakar, they will be transferred to internment camps to wait out the war. We can do no more and no less."

"And we Italians, Lieutenant?"

"Are on the right side in this wretched war. You have the freedom of the ship as long as you stay out of the way of the crew."

"And I can visit Stanislaw?"

"As I said, you have the freedom of the ship. Tell me your name again, Sergeant?"

"Scarpetti, Marco, Lieutenant."

"Is you English as good as your French, Scarpetti?"

"Better, Sir!"

The Frenchman looked askance at the diminutive Italian. *As if anything can be better than French!* Finally, he shrugged and said, "Very well. Stay by me, Sergeant, if you will. You will be of great service as an interpreter."

The Laconia Incident

"*Avec plaisir, Monsieur.*"

* * * * *

Later on, that same Thursday, the *Gloire* came upon the lifeboat that held the Logans and Tom Buckingham. The lifeboat's passengers were down to a cup of water a day, and a bit of biscuit. If baby Helen had not been breastfeeding, she would never have survived.

The French were solicitous of Violet Logan and her baby, and, by extension, of Donald as well, despite the remnants of his RAF uniform. They were far less solicitous of *Laconia's* third officer, until Marco spoke up for him.

"If it were not for Mr. Buckingham," Marco explained to the French lieutenant, "life aboard the *Laconia* would have been far more odious. It was Mr. Buckingham who insured we Italians received sufficient food and water, and got medical treatment when we needed it. He made sure we had adequate sanitation. If a bunk was available in the ship's sick bay, he gave it to an Italian. If it was not for him, fewer of us would have survived."

"Be that as it may," the lieutenant responded, "and as I said before, he will be treated honorably, but he is still technically the enemy."

"*Je comprends,*" Buckingham volunteered. "I understand."

"Ah, you speak French," the lieutenant noted in that same language. "Then you must know I have no choice. While the Vichy government is technically independent and neutral, we still answer to our German conquerors." The latter he said with a hint of sadness.

"*Encore, je comprends,*" Buckingham replied, in French. "I am, at least, alive, and once more with a solid deck under my feet. For that, *Monsieur*, I am most grateful."

11

Friday, 18 September, 1942, to Monday, 16 November

Over the next several days, beginning that Friday, 18 September, the Vichy cruiser *Gloire*, and the two sloops, continued rescuing those in lifeboats, and those few they found in the water who had managed to still stay alive.

But there was only so much the French could do, either for Buckingham, or for the others Marco had spoken for. The women and children were indeed treated very well, allotted accommodations in a section of the crew's quarters. The men among the English and the Poles, in contrast, were kept under guard in conditions no better than those the Italians had endured aboard the *Laconia*. Still, given the circumstances, even the men under guard would have admitted that the French treated them quite civilly.

* * * * *

On Saturday, 19 September, *U-156* approached and attacked the British freighter, *Quebec City*, sinking her with a single torpedo. Of her 45 crew members, 41 managed to get into two lifeboats and row away from their ship before she sank. It was just before four o'clock in the afternoon, when the *Quebec City* went down. The day was bright and sunny, with only a few billowy clouds. The gunmetal sea was calm, with only gentle swells.

The men in the lifeboats sighted a circling submarine's periscope and were quite frightened when the boat surfaced, some of its crew scrambling on deck to man its guns. Other crew members appeared, brandishing machine guns, still others manned the bridge as lookouts, scanning the skies and the horizon for any possible threat. Finally, *Korvettenkapitän* Werner Hartenstein appeared on the bridge. The *Quebec City's* Captain, William Thomas, described him only as a tall military figure with silver braid on his cap.

The men in the lifeboats were surprised when, in perfect English, Hartenstein instructed them to bring their lifeboats alongside. Once the boats were tied alongside, Hartenstein asked if the captain was aboard. Thomas stood up. "Captain William Thomas, Sir, late of the *Quebec City*, at your service."

"Ah, Captain Thomas, I am genuinely sorry to have sunk your fine ship. But such are the fortunes of war. Tell me, Sir, what did the *Quebec City* carry, and where were you headed?"

"To Freeport, Sir, with a cargo of Egyptian cotton."

"I see." Hartenstein paused. "I sincerely wish I could be of more assistance in assuring the survival of you and your crew, Captain, and I would have happily towed your boats close to land if I could—but I cannot. We were only recently attacked from the air by an American bomber while doing that very thing."

"Really, Sir?" Thomas said, registering his surprise.

"Most assuredly, Captain. While perhaps I cannot take you in tow, I can at least point you in the proper direction. As I see it, you have two alternatives. The closest land hereabouts is Ascension Island, about eight

The Laconia *Incident*

hundred miles south-southeast. The African coast is farther, some twelve hundred miles east. Your choice."

Captain Thomas considered his options for a minute. "Our chances of finding a tiny island with just a compass are slim, Captain. As I see it, our best choice is the African coast."

"I agree, Captain. Come aboard and you can look at the map—perhaps plan your voyage."

Thomas boarded the submarine and did just that. Later, the two lifeboats were left on their own, as the submarine sailed off to the northwest on the surface. Before leaving them, Hartenstein apologized for being unable to give them any food, but did give them some tins of water.

As they proceeded east, the two lifeboats became separated. One was discovered by a passing British destroyer after fourteen days at sea. Captain Thomas' boat made it to landfall in Liberia.

Given his dealings with the crew of the *Quebec City*, it was clear to Leopold Schumacher, and the other officers and crew aboard *U-156*, that their captain fully intended to observe Dönitz' *Laconia* Order by simply ignoring it.

* * * * *

Some thousand survivors of the *Laconia* aboard the *Gloire* were eventually landed in Dakar, Senegal, on 24 September 1942.

* * * * *

Dönitz was called to attend a high-level staff meeting in Berlin; the meeting was held on 28 September, and *Der Fuehrer* himself was present.

After more mundane issues were discussed, Hitler took Dönitz aside. First, he complimented him on the job the U-boats were doing in destroying Allied shipping in the Atlantic and disrupting England's vital supply lines.

"But there is another matter," he said. "Regarding this most recent incident off the African coast. You must drum it into your commanders' heads that there is no place for sentimentality in warfare! War is a cruel and unforgiving business, and the coddling of enemy survivors at sea is foolish and unproductive. It is utter nonsense to offer enemy survivors food, water, or sailing directions. I read your order to your U-boat commanders, and it is fine as far as it goes, but it does not go near far enough!"

Hitler was working up a head of steam. "You must tell your U-boat commanders that they are to shoot all survivors from sunken enemy ships—including those that are in lifeboats!"

"But *Mein Fuehrer*," Dönitz countered, keeping his voice as calm and steady as possible. "Are you at all sure that such a command would be wise? Please try to understand, *Fuehrer*, your U-boats are manned by volunteers, sailors steeped in maritime tradition. This is an honorable war they are fighting, and they fight for their *Fuehrer* and for their Fatherland. There is no honor in killing a defeated enemy fighting for his mere survival in a lifeboat. Please, *Fuehrer,* sending out such an order would utterly destroy the morale of your U-boat crews!"

Hitler paused and considered the words of his Commander of U-Boats. Of all Germany's fighting forces, land or sea, there was no arguing that his U-boat force was by far the most effective. *Perhaps,* he thought, *Dönitz might just be right. The very last thing I want to do is destroy morale in the U-boats.*

"Very well, then," he said. "Have it your way. It may well be that what you have already ordered is sufficient. But remember that we almost lost a U-boat because one of your commanders was far too chivalrous. I am well aware that Hartenstein is a hero of the Reich, but he went too far. Such behavior must not happen again!"

"Yes, *Mein Fuehrer,*" a relieved Dönitz replied. "It must not and it will not." But even as he spoke, he was thinking *And perhaps you understand after all, Mein Fuehrer, that submarine crews are not at all like your SS battalions.*

* * * * *

At least three of the lifeboats from the *Laconia* were not rescued by the French. One lifeboat, with Robby Cotton aboard, was rescued by the *Empire Haven.*

A second was picked up by a British trawler after forty days at sea. Of the fifty-two passengers originally aboard, one was James Fellows. Only four had survived, and Fellows was not among them.

Yet another was more fortunate, but only in comparison. The lifeboat in which Jim McLoughlin rode had set off toward the African coast, some 800 miles to the northeast. Of the sixty-eight passengers originally aboard, only sixteen had survived when it landed on the Ivory Coast some twenty-six days later, on 8 October 1942. Jim had celebrated (although that word hardly applied) his twenty-first birthday just six days earlier.

The Liberian natives who greeted them, feted them in their village overnight before putting them in dugout canoes the next day for transport thirty-five miles upriver to the relative civilization in Grand Bassa. It was while they were in Grand Bassa that one of the survivors developed sepsis in a leg sore from which he eventually died. Jim and the remaining fifteen survivors remained in Grand Bassa, and were cared for until they were well enough to be transported to Freetown; by then it was well past mid-October.

Sometime between his stay in Grand Bassa and his continued recovery in Freetown, McLoughlin contracted malaria, a disease that was to stay with him for the rest of his life. When sufficiently recovered to be considered fit to return for duty, Jim was drafted to *HMS Hecla*, a destroyer bound for England. McLoughlin couldn't believe his good fortune! But just as he was to be mustered aboard her, he was hit with a severe bout of malaria and became so sick that *Hecla* left Freetown without him. Now Jim couldn't believe his *bad* luck.

It wasn't until he was on his way to recovery several days later, that McLoughlin learned that on 12 November, the *Hecla* had been torpedoed by a German submarine in the Atlantic, west of Gibraltar, resulting in 281 casualties.

*****

On Monday, 16 November, 1942, Werner Hartenstein, in *U-156*, returned to port in Lorient, his command having completed her fourth war patrol.

Gene Masters

12

Robby Cotton: Friday, 18 September, 1942, to 19 January, 1944

It was on Friday, 18 September, when the lifeboat which carried Robby Cotton was discovered by the *Empire Haven*. The ship had only just reached the scene of the *Laconia* sinking, and was searching for survivors.

Robby, salt-encrusted and dehydrated, was barely able to acknowledge his rescuer's greeting. He learned that he had been promoted to Able Seaman when the *Empire Haven* made port in Liverpool on 29 September.

After a month's rest, he was drafted to *HMS Violet*, a corvette that made the regular run between Dover and Halifax, Nova Scotia, escorting convoys. The *Violet* was a Flower-class corvette, just 280 feet long, with a beam of 33 feet. Her top speed was 16 knots, which was just adequate for keeping station on convoys.

Violet was equipped with a type 271 SW2C air-and-surface search radar, and a type 144 sonar. She had only limited anti-air capabilities—her mission, after all, was anti-submarine warfare. The *Violet* was designed and equipped to detect, locate, and attack submarines with guns and depth charges.

Broad in the beam and shallow-drafted, the Flower-class ships were known both for their sea-worthiness and their rough ride. Even the saltiest seamen fought seasickness aboard the Flower-class corvettes, which were reputed to "bobble on a grass sea." On the other hand, these ships were also responsive to the helm, and capable even in the roughest of seas.

By 1943, the "happy time" in the Atlantic for Dönitz' U-boats was over, and U-boat losses began to mount. By then, the skies over the convoys were populated strictly by friendly aircraft, and the British had long since succeeded in successfully miniaturizing radar sets so that they could be mounted in those aircraft.

U-boat tactics had also long since evolved so as to attack in numbers and in unison, called *Reudeltaktic* by the Germans, and "wolf packs" by the British and Americans. U-boats also preferred to attack on the surface, where the boats were faster and more maneuverable than when submerged. But now, airborne radar allowed Allied planes to locate surfaced U-boats quickly, and then attack. Frequently, the planes were upon the boats before they could submerge, and a well-placed bomb would blow the boat to pieces. Even when boats were able to dive away from the air threat, they were forced to stay submerged until the plane cleared the area, and, by then, a targeted convoy would have long since sailed past the danger.

To thwart detection from the air, and from ship-mounted radar, the *schnorchel,* or snorkel, a device designed to sip air from the surface for running the diesels while submerged, was being retrofitted to U-boats as quickly as possible. With the snorkel, the boat could almost match surface speeds submerged, and were far more difficult to detect with radar. But, as with the new type XXI U-Boat class—the Wonder Boat—it was too little and too late.

The Laconia *Incident*

* * * * *

On 19 January, 1944, *Violet* was on escort duty for the combined convoys OS65/KMS39 in the Northwest Atlantic, just southwest of Ireland. The weather was miserable: cold, stormy, and overcast. The sea was rough indeed, with pounding waves ranging as high as thirteen feet. *Violet* was bouncing on the surface like a cork, pitching and rolling.

The *Violet* had done her part, ever since the convoy had formed in Halifax, and gotten underway, detecting and chasing U-boats, attacking with both depth charges and guns. Among her smaller-caliber guns, *Violet* was armed with a single-mount, four-inch gun, and Robby, now promoted to Petty Officer Third Class, became her gun captain. Since the skies over the convoys were populated strictly by friendly aircraft, there were no air targets. The gun crews on the *Violet* made up for it by firing on U-boats at every opportunity.

U-641, a type VIIC U-boat, *Kapitänleutnant* (Lieutenant-Commander) Horst Rendtel commanding, was operating with the wolf pack *Rügen* on her fourth war patrol, when it came upon the convoys OS65/KMS39.

By the early afternoon, Rendtel was positioning his boat on the surface, under a frigid, sunless, and overcast sky, for a run on the convoys. At the same time, *Violet's* radar operator picked up an intermittent surface contact amid the sea return. The contact might not have been anything at all—perhaps just a wave-tip higher than the rest—or it may have been an enemy conning tower. The operator thought for a bit before reporting it, and possibly crying wolf. Then he decided that the prudent thing would be to report it.

He reported the "possible surface contact" to the bridge, and the ship's captain, Royal Naval Reserve Lieutenant Charles Stewart, decided to leave station on the convoy, close, and investigate, and so notified Escort Command. As *Violet* was en route to the contact's location, Stewart set general quarters, and Robby Cotton and his gun crew made to man the four-inch gun. *Violet* was bouncing so violently that it took a good five minutes before Robby and his gun team was able to reach, and limber, its weapon. From his post on the gun mount, Robby was able to observe the action as it unfolded.

Rendtel was running his approach on the convoys from the U-boat's bridge when *Violet* was sighted bearing down on the submarine. Rendtel immediately ordered his boat submerged. Aboard *Violet*, the "clear to fire" order came down from the bridge, and Robby tried to track and aim at the diving U-boat. But aiming, much less firing, from the bobbing gun platform at the lurching, and slowly disappearing, submarine, proved near impossible. Robby was unable to get a shot off before the boat completely disappeared beneath the roiling sea.

When *Violet* finally reached the place where the submarine had last been spotted, the boat had already been underwater for several minutes. Captain Stewart ordered his ship to slowly traverse the spot, using active sonar in an attempt to locate it. *Violet's* sonar operator searched, and Violet got lucky; the sonar man received a satisfying return ping on the third traverse. With the U-boat located, the corvette began the first of three depth-charge runs. Upon completion of the third run, *U-641* shot to the surface, mortally wounded, not 500 yards off *Violet's* starboard bow. Stewart immediately ordered his ship to back away from the U-boat, opening up the distance between them, lest the submarine suddenly get tossed toward, and possibly strike his ship.

The men aboard the *Violet* watched as bodies scrambled from the receding U-boat's conning tower and into a hastily-inflated rubber boat bobbing in the water. When only about a dozen or so men had made it successfully into the boat, the U-boat suddenly shot forward, then backward, spilling the others still trying to abandon their U-boat into the sea.

The German submariners in the rubber boat quickly paddled away from their stricken craft, lest they get sucked down with it. From his position on the gun mount, Robby watched as the bow of the submarine lifted up, and then out, of the water. *U-641* then slipped below the waves stern-first. By then, the inflatable was well away from the spot; in fact, Robby had difficulty locating it in the rough sea. And now, sea conditions appeared to be getting even worse.

"Aren't we gonna shoot at 'em?" one of the loaders, a regular seaman named Hawkins, called up to Robby.

"Whatever for?" Robby replied. "They ain't a threat to us, nor to the convoys any more. What would be the point?"

"But they're the enemy!" the loader retorted. "We should blow 'em out of the water!"

"Should we now?" Robby said, almost pensively. "Have we come to that, now, have we, Hawkins? Have we British lost all traces of our humanity?" But Hawkins only scowled in reply, his blood lust up and unsatisfied. In truth, Robby dreaded the thought that an order to shoot at the survivors might come down from the bridge.

And so Robby was glad when the order did come down to secure general quarters, and for a rescue party to form up on the main deck.

Captain Stewart then had *Violet* close in on the spot where the boat had gone down, hoping to rescue as many survivors as possible. But not a single man could be located in the ever-worsening weather. Any attempt to locate the rubber raft with radar proved just as fruitless. It was now clear that the entire crew of the U-boat was probably lost.

The *Violet* then bobbled into a sharp turn and headed back toward the convoys. Robby would have preferred that the captain had stayed in the area of the sinking longer, and had tried harder to rescue the U-boat survivors, but he understood that the *Violet's* mission was to protect the convoys, and he really couldn't fault his commander's decision to return to station. *After all*, he thought, *these U-boats traveled in packs, and there may well be another one close by, waiting to pounce.* But there were no more submarine attacks that day.

Robby was used to lying in his heaving bunk, and had long ago learned to position his body so as to not be tossed onto the deck, even in the roughest storm. Indeed, he had learned to fall fast asleep in any weather the sea gods could conjure up. But that night, off watch and in his bunk, Robby could only think of the poor men who had gone to their deaths that afternoon. He wondered if Werner Hartenstein would have given up on rescuing *Lavonia's* flotsam as quickly as *Violet's* commander had done with the U-boat. Of course, conditions in the South Atlantic had been entirely different then, and it wasn't really fair to compare the two events — or the two commanders, for that matter.

Robby wondered if Hartenstein and his crew would survive the war. If they did, he knew, it would mean that Hartenstein and *U-156* might well destroy yet more ships in the interim, kill even more of his countrymen and their allies before the last shot was fired. Still, he hoped that there was some way Hartenstein might yet survive, and they could meet, and he could thank him for sparing his life and the lives of the others who ultimately survived. Then Robby slept.

Gene Masters

Epilogue

September, 1942, and thereafter

The British civilians who had been aboard the *Gloire* were interned in an old French Foreign Legion camp. The English and the Polish servicemen were sent to Casablanca and interned in a French prisoner of war camp. Under orders from their Nazi overlords, the French were directed to keep the British and the Poles separate, and arrangements were to be made later to transport the Poles back to Poland, where they were to be interned in "work camps."

Marco Scarpetti was allowed to leave the *Gloire* along with its other Italian passengers when it docked, but the Italians soon found there was really no place for them to go. Italian government operatives in Senegal tried to arrange transport of its personnel back to Italy, first by ship to a southern French port, and then overland through France to Italy, but no shipping was available. Sending them back to their units in Italy and North Africa overland on African soil was another alternative. Italian authorities worked on both plans while Marco and the other rescued Italians languished in a makeshift tent camp through October.

German authorities, meanwhile, faced the same problems in their attempts to "repatriate" the Polish POWs to Poland. The Germans, however, had more access to available transportation facilities than did their Italian allies. They had set up an overland truck convoy to take the Poles to Algiers, there to be transferred to an Italian ship, which would land them in Messina, Sicily, and thence overland by rail back to Poland. But trucks for the initial leg of the journey would not be available until mid-October. Mid-October came and went, but the trucks did not. They apparently had been commandeered for a more important mission, and were now promised for early November.

The Allies launched Operation Torch, the invasion of North Africa, on 7 November 1942. Operation Torch interrupted the final preparations to send Stanislaw Kominsky and his fellow Polish prisoners back to Poland; the requisite trucks now were needed to ferry troops and supplies to the front. The Poles, therefore, remained where they were, in French custody, just outside Casablanca.

By 16 November, British and American forces had secured French North Africa, and the *Gloire* and the other French warships operating out of West and North Africa rejoined the Allies. In response, German forces invaded Vichy France starting from the Atlantic coast, and, by 11 November, German tanks had reached the Mediterranean coast. Italian forces, meanwhile, occupied the French Riviera, and had landed in Corsica.

The tent camp housing the Italians became, technically, a POW camp, and the rescued Italians once again became prisoners of war, this time French prisoners. But their French captors posted no guards, and meals were still regularly served. No one forced the Italians to stay in the camp, but there was really no other place for them to go. They busied themselves building and erecting more permanent shelters, and organizing football teams. Marco taught language classes.

The Laconia Incident

When Italy eventually capitulated that following September, 1943, the Italians in the camp were once again *former* POWs, but, once again, nothing much changed. Sicily was in Allied hands, as were parts of the southern mainland, but the bulk of the Italian peninsula itself was still in German, and die-hard Italian Fascist, hands. It wasn't until the final German capitulation, in May 1945, that the Italians were repatriated, and only then was Marco able to return to Bologna.

* * * * *

Shortly after the *Hecla* was sunk, in mid-November 1942, and with Jim McLoughlin's second bout with severe malaria ended, he was drafted to *HMS Dragon*, an aged cruiser. *Dragon* eventually brought him back from Freetown to his home in Liverpool in time to celebrate Christmas with his family.

The following June, now posted to *HMS Drake* in Plymouth on extended leave, he met a Royal Army Wren, Dorothy Field, who, at the time, was just nineteen.

When Jim McLoughlin's extended leave in Plymouth ended, he was drafted to *HMS Implacable*, a newly commissioned carrier, fresh out of the shipyards. Before reporting aboard, however, on 12 September 1944, he and Dorothy were married. Aboard *Implacable*, Jim once again was hit with a bout of malaria—severe enough that he had to be detached from the ship and treated ashore.

Recovered yet again, Jim was drafted to *HMS Golden Hind*. But the *Golden Hind* was in Sydney, Australia.

McLoughlin was unhappy to be separated so far from his new, and now pregnant, Dorothy, but he dutifully made his way to Australia to join his ship.

Once there, McLoughlin fell in love with Australia and her people, and, after the war, he and Dorothy sailed to Australia and settled there, and it was there that they remained permanently, and there they raised their family.

Whenever questioned about the *Laconia* incident, McLoughlin always admitted that he had been favorably impressed by Hartenstein, and was grateful for his kindness and gentlemanly behavior.

* * * * *

On 8 March 1943, fifty-two days into her fifth war patrol, and with Hartenstein still in command, *U-156* was attacked from the air and sunk by four depth charges released from an American PBY Catalina. The attack occurred just east the Island of Barbados, in the Atlantic. There were no survivors.

Similarly, two months before *U-156* met her fate, *U-507* was attacked and sunk by another PBY off the Brazilian coast. Harro Schacht went down with his entire crew.

In July, 1943, *U-506*, sailing off southwest Spain, was attacked and sunk by a B-24 Liberator bomber. There were six survivors; Erich Würdermann was not among them.

The Laconia *Incident*

* * * * *

The Armistice of Cassibile was signed on 3 September 1943, ending hostilities between the Kingdom of Italy and the Allies. A few days later, Luigi de la Penne was offered the opportunity to be released from prison by the British, provided he join the fight on the side of the Allies. De la Penne had no love for the Nazis. He accepted.

De la Penne went on to distinguish himself in a joint Italian/British operation against the Germans at La Spezia in 1944. After the war, he stayed in the Italian Navy, rising to the rank of vice admiral. A class of Italian frigates was later named after him.

* * * * *

The *Commandante Cappellini* was converted by the Italians into a cargo carrier and blockade runner, and renamed *Acquilla III*. She was sent to the Indian Ocean to smuggle material for the German and Japanese forces operating there. After the Italian capitulation, she was seized by the Germans, and rechristened *UIT-24*. Then again, with the defeat of Nazi Germany, she was appropriated by the Imperial Japanese Navy, and received yet another new designation, *I-503*.

When Japan capitulated, the U.S. Navy towed *I-503* out to sea and sunk her. Thus it was that the *Cappellini* had the distinction of having seen service under all three Axis powers.

* * * * *

After the liberation of France, in late August 1944, the civilian prisoners held in an old French Foreign Legion Camp in North Africa were released. Among the first to be released was Violet Logan and her now toddler daughter, Helen, who made their way to Wales, and to her husband's family in Ynysybwl. They arrived that Christmas.

For whatever reason, the British servicemen were not immediately released. So it was that Donald Logan was unable to follow Violet to Wales until months later.

Whenever asked, the Logans always stressed that their survival from the sunken *Laconia* was the result of the humane qualities exhibited by the U-boat commander, Werner Hartenstein.

* * * * *

Also released from the Internment Camp in Casablanca, in late November, 1944 was Stanislaw Kominsky. Upon their release, the Poles were instructed to arrange with the British authorities for transportation back to Poland. For Stanislaw, that would have meant return to military service. But Poland was, at the time, under the protection of the Soviet Union, and Stanislaw did not relish the thought of serving in a Polish army answering to Russian overlords; it was the Soviets, after all, who had raped his country, along with the Nazis, in 1939.

So Stanislaw never bothered to report to the British for repatriation, and decided instead to wait it out until the situation in Poland stabilized. Only then would he

The Laconia *Incident*

quietly make his way back to Poland on his own schedule. In the internment camp, he had developed passable French, so, he decided to settle somewhere in France for the interim. "Somewhere in France" became southwest France near the town of Auch, in Gascony. There, he obtained work as a laborer on a farm. The situation was, in Stanislaw's mind, only temporary.

A year passed, and the situation in Poland only grew worse from Stanislaw's perspective: the Soviet Union now exerted complete control over his country, and a puppet Polish Communist government had been put solidly in place. Meanwhile, he had met a young lady, the daughter of the owner of the farm on which he was working.

Kominsky family descendants can still be found in Gascony.

* * * * *

The military career of Captain Robert Richardson was not in the least way adversely affected by the *Laconia* incident. Richardson rose through the ranks, retiring as a United States Air Force Brigadier General in 1967.

* * * * *

At the end of the war in May 1945, before committing suicide, Adolf Hitler turned over the reins of the German government to Karl Dönitz. As the sole ruler of Nazi Germany, it was Dönitz who accepted the Allied surrender terms.

The Nuremburg War Crimes Trials followed the war, with indictments handed down by the four-power International Military Tribunal. Donitz was among those indicted. The tribunal had a very difficult time showing that Dönitz had behaved in any way contrary to the rules of war.

In the end, they pointed to the *Laconia* Order as proof that, in issuing the order, Dönitz had, in fact, directed his submarine commanders to behave contrary to the rules of war. In so doing, the tribunal ignored the fact that American submarines in the Pacific had waged unconditional submarine warfare against Japan throughout the entire war. Karl Dönitz served eleven years in Spandau prison before being released in 1956.

* * * * *

Robby Cotton left the *Violet*, and the Royal Navy, in September, 1945. He returned to Scotland, to Dalmuir, and even took a job as an assistant foreman at the William Beardmore and Company Shipbuilding Works, the very same shipyard where his father had worked. For a while he seemed content to work day-to-day and return to a flat he rented in Dalmuir, not too far from where his boyhood home once stood (now just open land cleared of rubble). He had even taken up with a young woman who had been four classes behind him at St. Stephans.

But Robby grew restless. He left Dalmuir, and made his way to Liverpool, where he took a job as an ordinary seaman aboard a merchant vessel. He crisscrossed the oceans several times before leaving the merchant service in 1953. He returned to Scotland, where he entered Prinknash Abbey, in central Scotland, some forty miles north of Loch Lomond. Robby took final vows as a Benedictine Monk at Prinknash just four years later. He died there in 1998.

Gene Masters

Acknowledgments

This novel has been written to coincide as closely as possible to the actual chain of events as described. The following reference materials have been invaluable:

The Sinking of the Laconia and the U-Boat War, James P. Duffy, University of Nebraska Press, 2009.

One Common Enemy, Jim McLoughlin with David Gibb, Wakefield Press, 2006.

http://www.uboataces.com/uboat-type-vii.shtml   Type VII U-Boat

https://www.uboat.net/articles/index.html?article=33   The *Laconia* Incident, Gudmundur Helgason

https://www.gingkoedizioni.it/the-sinking-of-the-laconia-not-always-bad-and-good-are-on-one-side/   *The Sinking of the Laconia – Not Always Good and Evil Are on One Side*, Angelo Paratico, 29 Oct 2013.

https://www.scotsman.com/news-2-15012/medal-honours-the-heroics-of-wartime-ships-captain-1-761584   *Medal Honors Heroics of Wartime Ship's Captain*, The Scotsman (newspaper), 23 July 2009

https://www.walesonline.co.uk/lifestyle/showbiz/baby-helen-story-laconia-1891494   *Baby Helen and the Story of the* Laconia, 28 Mar 2013.

https://www.thefreelibrary.com/%27I+survived+German+torpedo+attack...%27.-a0168616643 *I Survived German Torpedo Attack,* 11 Sept 2007.

https://uboat.net/men/hartenstein.htm *Top U-Boat Aces: Werner Hartenstein*

http://wernerhartenstein.tripod.com/U156Laconia.htm *Werner Hartenstein and the* Laconia *Incident*

http://www.sath.org.uk/edscot/www.educationscotland.gov.uk/scotlandshistory/20thand21stcenturies/worldwarII/clydebankblitz/index.html *The Clydebank Blitz*

https://www.scotsman.com/regions/glasgow-strathclyde/recalling-the-clydebank-blitz-76-years-on-1-4391237 *Recalling the Clydebank Blitz – 76 Years On,* The Scotsman (newspaper), 13 Mar 2017,

https://www.awm.gov.au/collection/C11676 *Crew of the* HMT Laconia *During a Gun-Drill*

https://www.bbc.com/news/uk-scotland-33092351 Lancanstria: *The Forgotten Tragedy of World War Two,* Graham Fraser, 13 June 2015.

About the Author

Gene Masters is a retired consulting engineer living in East Tennessee with his wife, Ruth. They have two grown daughters, and two grandchildren. He is the author of several technical treatises, including his doctoral dissertation. *The Laconia Incident* is his third work of fiction.

Masters received a commission in the U.S. Navy on graduation from Notre Dame, and his first tour of duty was aboard a transport in the Western Pacific. His second tour was aboard a recommissioned and updated diesel-electric submarine, the USS Angler. Angler was originally commissioned in 1943, and made seven war patrols in the Pacific before being decommissioned. Her updating to an SSK-class boat in the 1950s fitted her for operation against cold war submarine adversaries with advanced soundproofing and sonar. Masters left Angler and active duty after a Mediterranean tour. Later Naval Reserve assignments included the diesel-electric submarines USS Manta and the USS Ling.

After active duty, Masters pursued a career in engineering, and served in various companies until settling into a career as a consulting engineer. He retired in 2009. Readers interested in learning more about the author can visit his website at: www.genemasters.net.

Gene Masters

www.ingramcontent.com/pod-product-compliance
Lightning Source LLC
Chambersburg PA
CBHW022113040426
42450CB00006B/681